IMAGES
of America

BUILDING THE ASHOKAN RESERVOIR

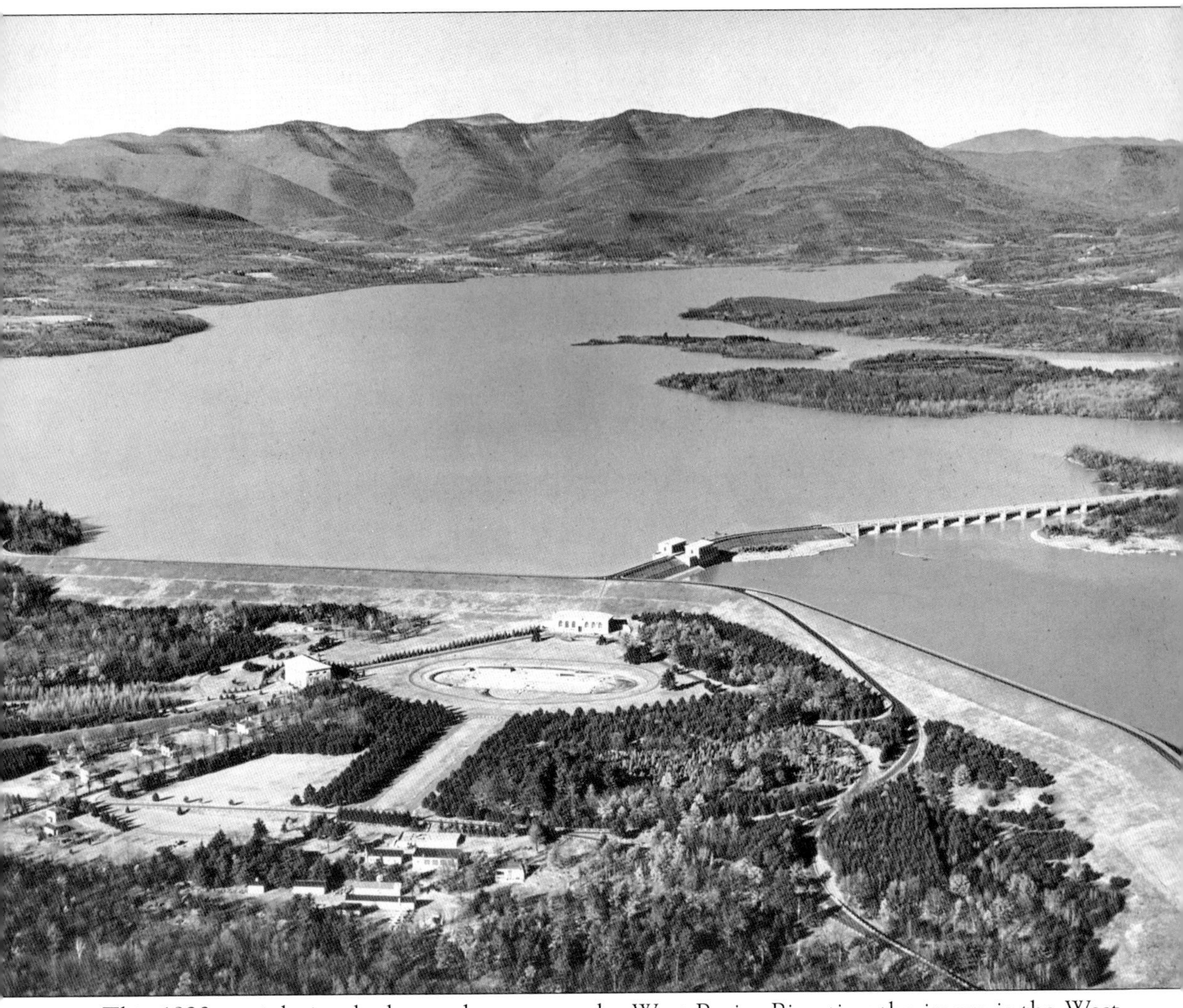

This 1920s aerial view looks northwest over the West Basin. Bisecting the image is the West Dike and the dividing weir. The aerators, screen chamber, and operator quarters are in the left foreground. Above the dikes is the Catskill mountain range. The Middle Dike runs to the right. Both dikes are reserved for pedestrians. Cars are allowed only on the dividing weir and down the Middle Dike road. (Courtesy of the New York State Archives Partnership Trust.)

On the Cover: This image illustrates the area's geologic history and the reason this location was selected for a dam. The horizontal strata of the Devonian period are covered by many feet of glacial lake sediment. Here, many deeply excavated rock layers are being readied for the foundation of a mile-wide dam that will make a nearly 123-billion-gallon reservoir. The human energy and sacrifice necessary to build this dam in seven years were enormous. (Author's collection.)

IMAGES
of America

BUILDING THE ASHOKAN RESERVOIR

Frank Almquist
Foreword by Kate McGloughlin

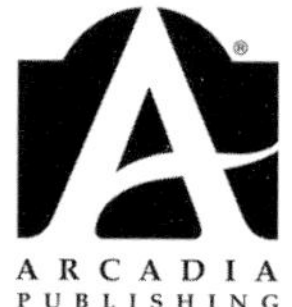

ISBN 978-1-4671-0725-9

Published by Arcadia Publishing
Charleston, South Carolina

Printed in the United States of America

Library of Congress Control Number: 2021943557

For all general information, please contact Arcadia Publishing:
Telephone 843-853-2070
Fax 843-853-0044
E-mail sales@arcadiapublishing.com
For customer service and orders:
Toll-Free 1-888-313-2665

Visit us on the Internet at www.arcadiapublishing.com

This book is dedicated to all who once called the Esopus Valley home and their descendants who may still carry the loss.

Contents

ACKNOWLEDGMENTS

Thanks to my wife for proofreading/editing and Kate McGloughlin for her magnificent foreword to this book. Then, in no particular order, thanks go to Ron Lamont, John Ham, David Turner, Audrey Klinkenberg, the Hurley Historical Society, New York Public Library (NYPL) Digital Archives, Ulster County Archives, Town of Olive Archives, Jeff Almquist, New York State (NYS) Archives Partnership Trust, American Society of Civil Engineers Archives, New York City Department of Environmental Protection, and others who answered many varied and sundry questions. All images not credited are from the author's collection of postcards and other material.

FOREWORD

The astonishing beauty of the Ashokan Reservoir and its daily gift of inspiration and familiar comfort belie a sorrowful narrative that seems to sigh beneath its surface.

What remains of the visible landscape—eroded plateaus that slope to the water's edge, a handmade shoreline that beckons both eagles and herons—holds a story that is a classic depiction of the symbiotic relationship between natural history and civilization, the migration, approach, and retreat of each leaving lasting physical and emotional vestiges, some of which are illustrated on the following pages.

The evolutionary record—evidenced by traces of Ice Age comings and goings, the birth of landmasses and swamps, the shaping of mountains and their tributaries—sets the stage that may inform our opinion on the story. Layers of time and the sway of both a natural evolution and human footprint have hammered, honed, and put a silvery polish on the view that is ours for now. Over millennia, the forces of nature and humanity have conjured a magnificent record of the fruits of this struggle.

After the glaciers retreated, deposits of fertile soil, breathtaking gorges, bluestone quarries, and rock-slicing, fished-filled streams attracted the first people known to this "place of many fishes." The Leni Lenape, following water and food sources, lived sustainably on this land, which was once a glacial lake, for generations. Their presence had little deleterious impact on this idyllic valley; their history of loss nearly forgotten even as the names they left behind identify landmarks and settlements in the townships that, to this day, embrace the Ashokan Reservoir.

Following water—it's what humans do. The pictorial story on the following pages tells of this civilization-old impulse of using natural resources to compete for still other natural resources. Humans who became known as "white settlers" came across the Atlantic Ocean, entered the mouth of the Hudson, and eventually found their way, using the Esopus Creek as their guide, to what is now the site of the Ashokan Reservoir in the Catskill Mountains of Upstate New York.

Their approach signaled the ultimate dislocation of the Leni Lenape, as waves of white settlers, land grants from the British Empire, and ensuing wars and subsequent broken treaties made a peaceable life there nearly impossible. This unfathomable rupture to their communal, agrarian lifestyle, this disconnection from the life-giving landscape they knew, marks a pivotal moment in the Esopus Valley's history. The concept of land ownership had replaced the tradition of land use.

By the early 1700s, outposts had become homesteads, settled land was granted ownership, and farmers wrestled stones from their fields as tanners, colliers, and quarrymen made use of the once ubiquitous resources to foster industries that later made the Catskills famous. Profitable industry made transport necessary, and by the end of the Civil War, railroad ties that were set to carry goods to market became the means by which the next wave of industry to claim the Catskills—tourism—could find its way.

While some people made their living by extracting from nature, others used her forces to secure their living for generations; some came and went while others stayed. In 1790, Asa Bishop began building a gristmill lured by the power of the water of the Esopus at a place later called Bishop Falls in Olive City. His family, like many others, planted fields of hay, wheat, and oat from the previously deforested lowlands of the Catskills and made his life and his living milling grains for his family and community. His son Jacob, known as "the Blind Miller," carried on this trade, as did Jacob's sons.

To walk the rutted roads through the 12 hamlets of the Esopus basin (Ashton, Boiceville, Broadheads Bridge, Brown's Station, Glenford, Olive, Olive Branch, Olivebridge, Olive City, Shokan, West Hurley, and West Shokan) at that time was to witness burgeoning communities. Churches and other public buildings had by then centralized townships that were loosely governed by public servants who might also have been the postmaster and owner of the general mercantile.

Families that relied on other families for community, in the highest sense of the word, became entrenched in civic and social structures. Durable relations with commonality in norms and ideals had, for the practical and spiritual support of small-town survival, come to fruition. At one time, there nearly existed a form of subsistence agricultural economy; blacksmiths, sawyers, and woodworkers offered timber, tools, and furniture to builders; builders supplied farmers; farmers supplied millers; millers supplied families with grains and seamstresses and tailors with woolens and broadcloth; and seamstresses and tailors provided clothing for all.

The interdependence of the Esopus Valley's citizenry cannot be overstated. Recorded minutes from church meetings and written personal diaries swell with tales of community suppers held on snowy winter nights where attendees arrived by horse and sleigh and of quilting bees and bake sales that raised money to build parsonages. There existed benevolent societies including the International Order of Odd Fellows and the Knights of Pythias that contributed to the well-being of their neighbors in need; well before forms of insurance and social and rescue services were developed, these institutions, chartered by family members and friends, found ways to streamline avenues of aid to those in need.

According to written accounts, it was not unusual to change the schedule of one's day to help right an overturned wagon or catch a runaway team of horses. Teamsters carrying lumber or bluestone or driving sheep to Kingston often stopped en route to speak with passengers in smaller wagons, relaying messages and news along the once planked, now bluestone thoroughfares. The sound of shod horses clapping on bluestone tracks echoed the clang of hammer on anvil emanating from Mose Palen's blacksmith shop. Church bells that called the faithful on Sunday mornings were also used to signal a need for community response during a barn fire or farming accident.

By the mid- to late 19th century, the Catskills in general and the Esopus Valley in particular had become a destination for city dwellers seeking refuge from crowds, summer heat, and the undesirable conditions arising from crowded urban life. Boardinghouses and hotels swelled, as did shops and restaurants. Many homes became country retreats and second homes to New Yorkers who had fallen under the spell of the arresting landscape and bucolic lifestyle of the valley. Natural beauty had become another natural resource.

Just miles away, the majestic Hudson River, which had been bringing European settlers since 1604, seemingly sewed these mountains to the sea. Settlements from Manhattan to Kingston sprung up and added to the transactional and relational alliances being formed in the greater region of what is now known as the Mid-Hudson Valley.

Water had become both the means and the end in our story.

In the late 19th century, engineers were sent in search of a solution to Manhattan's age-old problem: providing clean drinking water to its ever-growing populace. Immigrants seeking refuge were crossing the Atlantic in record numbers, adding to the city's growing need for water. Though the Hudson pours next to the island of Manhattan and land on the east side of the Hudson might have made for a straighter pipeline, like the Leni Lenape and the likes of Asa Bishop, the city of New York's Board of Water Supply (BWS) chose the Esopus Valley and the power of the Esopus Creek at Bishop Falls to meet its needs.

As early as 1895, the Ramapo Water Company determined that the Esopus Valley was the ideal location to build a dam that would hold enough water to bolster the existing storage at the

Croton Reservoir and supply New York City with a seemingly endless supply of pure and wholesome water. This company's scurrilous attempt at land grabs to secure water rights that would enable it to sell water to New York City at a prohibitive cost were summarily exposed and deterred by public outcry and careful scrutiny by one commissioner and the sitting comptroller. A subsequent state water commission and a succeeding board of water supply appointed engineers to test the viability of the Esopus Valley. Waldo J. Smith, an impressive young engineer, was hired, and the urban normative sentiment prevailed: "Rural communities must be sacrificed for the needs of the great city."

Within five years, public hearings were taking place, and boring tests were being taken at Bishop Falls before the state water commission had even approved the project.

Farms and orchards, graveyards, schoolyards, and swimming holes, tended to and reveled in by generations, were being surveyed with a disinterested eye, soon to be dismantled, razed, and nearly erased from history.

The tearing apart of these communities, an ironic 200-year-old echo of the deep wound experienced by the Leni Lenape, was also the shredding of ties between rural and urban, city and country dwellers. Reportedly, a few residents were happy to leave, and some were paid "handsomely" though well below real market value, but most were dispossessed, forced to move upon 10 days' notice, and paid a fraction of what might have been fair. A property with a rightful valuation of $5,000 might be assessed at $500, and the owners finally paid $250. A number of claimants died before their claim was settled and disbursed.

They exhumed their dead—with some remains relocated to upland private cemeteries and the unknown disseminated to a patch of land in a hollow in the shadow of High Point Mountain, now called the Bushkill Cemetery—families dispersed, and the loss of livelihood and opportunity became a burdensome heirloom left to future generations that remained at the shoreline of the former Esopus Valley.

An undeniable feat of engineering and need for clean water for what was at that time the most important city in the world had replaced an idyllic life for those who settled near that clear water to build their life. Perhaps an accelerated scraping of evolution, certainly a competition of natural resources, though humanity has inexorably altered the landscape that was once here, the persistence of the Esopus Creek continues to sustain life, now quenching the thirst of upwards of nine million souls daily.

Kate McGloughlin
Olivebridge, New York

The following resources were consulted in the writing of this foreword:

Arold, Eleanor Ruth Whiting. *A History of Glenford.*

Bierhorst, John. *The Ashokan Catskills: A Natural History.* Fleischmanns, NY: Purple Mountain Press, 1995.

Davis, Elwyn. The Diary of Elwyn Davis 1906–1909, 2017.

Lee, Du Mond Frank. *Walking through Yesterday in Old West Hurley.* Kalamazoo, MI: Different Dimensions, 1990.

Rowe, Allen M. *Old West Hurley Revisited: A Nostalgic Tour.* Saugerties, NY: Hope Farm Press, 1999.

Steuding, Bob. *The Last of the Handmade Dams: The Story of the Ashokan Reservoir.* Fleischmanns, NY: Purple Mountain Press, 1985.

Weidner, Charles H. *Water for a City.* New Brunswick, NJ: Rutgers University Press, 1974.

INTRODUCTION

This book started with a collection of real-photo postcards I amassed over the years. An engineer by profession and fly fisher of the Esopus, I was led to learn more about the reservoir. Over time, acquiring more cards and books, my collection began to take on a life of its own. By themselves, they were interesting, but together they told a visual story—all that was needed was the written word to fill in the blanks.

There is engineering and design to discuss. This book looks at the "how" and not the "why." The ability to convert inked lines on drafting paper into a 200-foot tall structure takes special skills this book can only picture. The manpower needed to build such structures by hand (using air- and steam-powered equipment) was beyond ordinary. Wooden forms nailed together produced shapes that allowed water to flow smoothly for the past 100 years and with more years to go. Work on the dam was typically from mid-March to mid-December.

With the varied disciplines and skills used to build diverse sections of the dam, this book takes a different tack to look at the pieces. After a brief introduction to the valley communities, the Olive Bridge dam, weir, and dike construction are described. Then, a snapshot of the people who were there for the duration and where and how they lived is described. Their tools are investigated. In the days before the internal combustion engine and computers, hands were the tools. A few images in the last chapter show the bridges and reservoir today, finishing the journey.

Ulster County courts were the dispute mediator with the New York City commissions over resident land payments, and records were public. The county clerk's office made that information available at the Town of Olive Library on Route 28A. The library is becoming a repository of other Ashokan materials.

There is a lot left out of this book and a lot more to be found in other books. The bibliography lists those used here. All have extensive references that can lead the reader into much greater detail. One can dig as deeply as they care to into all parts of the story.

Throughout this book, frequent reference is made to *The Catskill Water Supply of New York City* by BWS engineer Lazarus White, a concise study of the reservoir's design and construction.

One

The Valley before the Reservoir

In the late 1890s, New York City had a water famine. The population was increasing by nearly 1,000 people weekly, and existing water supplies were inadequate. In the early 1900s, the Board of Water Supply started looking to the Catskill watershed. In 1904, the state legislature investigated land acquisition. On Saturday, August 5, 1905, state water commissioners spent several days in a conference on the valley at the Grand Hotel in Highmount. They commented that with the valley's prosperous appearance and population density, a reservoir would come with a high cost. On May 14, 1906, the legislature approved taking the land.

The valley was mapped, with all parcels surveyed and photographed. Everyone had to be off the land by September 1913. West Hurley, West Shokan, Boiceville, and Shokan relocated their post offices and hamlets. Ashton, Olive City, Browns Station, and Broadhead's Bridge were abandoned. Buildings were disassembled or moved intact. Buildings not moved were burned. Thirteen miles of Ulster & Delaware Railroad (U&D) mainline track had to be relocated outside the reservoir area—without disturbing service.

In *Water for a City*, Charles Weidner writes,

> In the area taken by New York City, there were 504 buildings, three times that number of barns, shops and outbuildings, 9 blacksmith shops, 35 stores, 10 churches, a grist mill, and 7 sawmills. There were general stores, meat markets, millinery and watch repair shops, a print shop, photographer's studio, harness and wagon shops, an excelsior mill, one large and several small stone "docks" where bluestone was cut and loaded, a creamery, a heading mill, barber and cobbler shops, bowling alleys and recreational halls. Living in the valley were craftsmen and professional men to perform practically every service required by the inhabitants.

Very few residents received fair value for their businesses and properties, leading to many Ulster County court cases.

The following pages provide images and descriptions of the communities in the same order they would have been encountered aboard a westbound Ulster & Delaware Flyer.

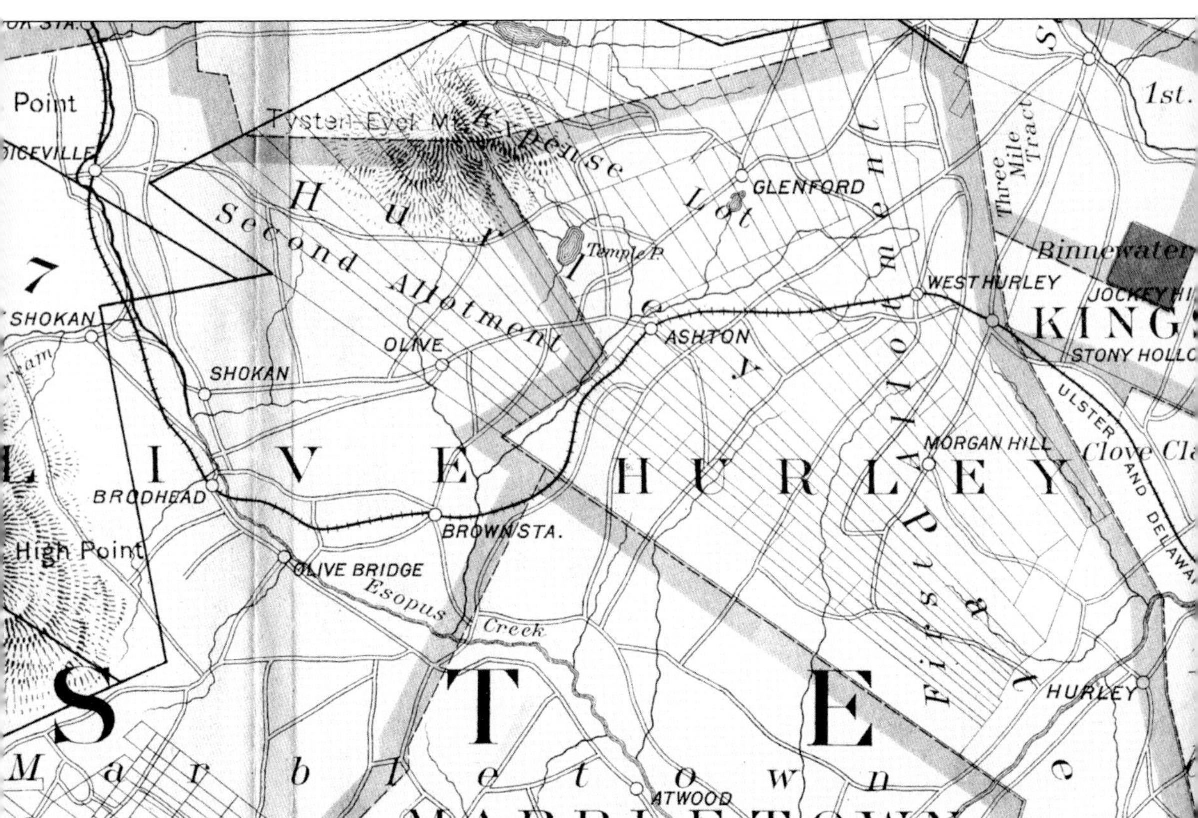

This portion of a late 1800s map of the central Hudson River valley illustrates the Esopus Valley. The U&D Railroad route starting near Kingston meanders through the valley. Hamlets, roads, the Esopus, and High Point are all on the map and in many of the images that follow.

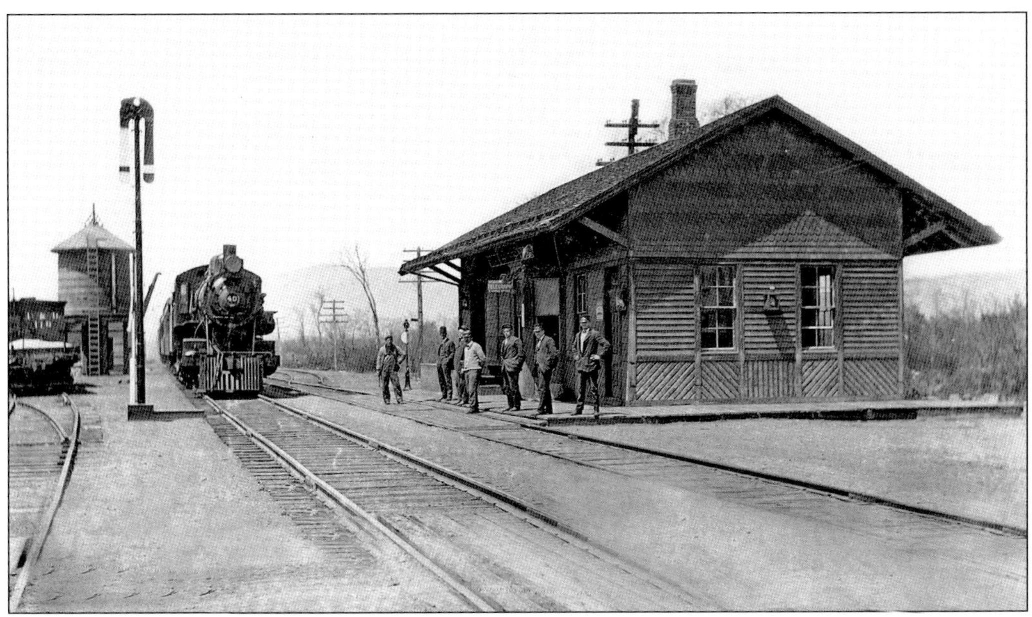

The West Hurley station sat on a hill a short distance north of the hamlet center on Railroad Avenue. Here, a U&D train is heading east into Kingston to stop for a few passengers. The ticket price for a trip from West Hurley to Kingston was 39¢. (Courtesy of John Ham.)

MAIN LINE—ONEONTA TO KINGSTON POINT

STATIONS	EAST BOUND FIRST CLASS											Minimum time between Stations-Freight Trains	Train Order Offices.	Water Stations.
	30	8	32	34	14	36	28	18	38	4				
	Mount'n Express	Day Line	New York Express	Catskill Mount'n Limited	Local	Rip Van Winkle Flyer	Milk	New York Local	Sunday Mount'n Special	Local				
	Daily Except Sunday	Daily Except Sunday	Daily Except Sunday	Daily Except Sunday	Sunday only	Daily	Daily	Daily Except Sunday	Sunday only	Daily Except Sunday				
LEAVE	A M.	A.M.	A.M.	A M.	P.M.	P.M	A.M.	P.M.	P.M.	P.M.				
Phoenicia A	s 7 46	s 10 31	s 10 57	s 1 12	P.M.	s 4 18	s 5 08	s 6 26	s 7 42	....	....	14	N	W
Brodhead's Bridge ...	8 15	s 11 07	s 11 28	1 35	s 4 27	4 48	5 43	s 7 00	8 14				.	
Brown's Station......	s 8 20	s 11 13	s 11 34	1 39	s 4 33	4 53	s 5 49	s p 7 08	8 18				.	
Olive Branch.........	8 25	s 11 19	s 11 40	1 43	s 4 39	4 58	5 54	s 7 14	8 22				.	
West Hurley.........	s 8 31	s 11 25	s 11 46	1 47	s 4 45	5 03	5 59	s 7 20	s 8 29				.	
Stony Hollow........	8 35	s 11 30	11 50	1 50	s 4 50	5 06	6 03	s 7 25	8 33				.	
Kingston.......... A	s 8 45	s 11 40	s 12 00	2 00	s 5 00	s 5 18	s 6 15	s 7 35	s 8 45				.	W
Kingston.......... L	8 45	11 42	12 00	2 00	5 05	5 18	j 6 15	7 35	8 45				.	
Rondout............	A 8 55	s 11 50	A 12 10	A 2 10	A 5 15	A 5 28	A 6 25	A 7 45	A 8 55	....	....	10	N	W
Kingston Point......	....	11 55	....	....	....	....	....	....	....	....	....	5		
ARRIVE	A.M.	A.M.	P.M	P.M.	P.M.	P.M.	P.M.	P.M.	P.M	P.M				

This U&D eastbound timetable shows the morning train stopping at West Hurley at 8:31 and arriving in Kingston at 8:45—a 14-minute journey, all downhill. Travel time between stations was short. Tickets were probably available for travel from West Hurley to Brown Station or other nearby hamlets. The roads were dirt and only good in dry weather.

MAIN LINE—KINGSTON POINT TO ONEONTA

STATIONS	WEST BOUND FIRST CLASS											Distance between Stations.	Distance from Kingston Point.
	3	27	9	31	33	7	35	37					
	Local	Milk	Local	Catskill Mount'n Limited	Mount'n Express	Day Line	Rip Van Winkle Flyer	Ulster Express					
	Daily Except Sunday	Daily	Daily	Daily Except Sunday	Daily	Daily Except Sunday	Daily Except Sunday	Daily Except Sunday					
LEAVE	A.M.	A.M.	A.M.	P.M.	P.M.	P.M.	P.M.	P.M.					
Kingston Point	....	....	....	(See Note)	(See Note)	(See Note) 2 30	(See Note)	....	....	....	....	0.0	0
Rondout	....	5 35	6 45	12 15	1 50	2 35	3 30	6 05	....	....	....	1.0	1
Kingston A	....	s 5 43	s 6 53	12 23	s 2 00	2 45	s 3 38	s j 6 15					3
Kingston L	....	6 10	7 10	12 35	2 20	2 45	3 55	6 40					
													9
Stony Hollow	....	6 25	s 7 20	12 44	s 2 30	2 54	4 04	6 49					10
West Hurley	....	6 40	s 7 31	12 51	s 2 41	s 3 04	4 12	s 6 59					13
Olive Branch	....	6 46	s 7 37	12 55	s 2 47	s 3 10	4 16	7 04					16
													19
Brown's Station	....	6 51	s 7 43	12 59	s 2 53	s 3 16	4 20	p 7 08					20
Brodhead's Bridge	....	6 56	s 7 49	1 03	s 2 59	s 3 22	4 24	7 12					22
Shokan	....	6 58	s 7 54	1 05	s 3 04	s 3 27	4 26	s 7 16					23
													26
Phoenicia A	....	s 7 17	s 8 18	s 1 20	s 3 26	s 3 49	s 4 41	s 7 35	....	....	....	2.6	28
Phoenicia L	....	7 22	8 26	1 29	3 36	3 58	4 50	7 44	....	....	....		

This westbound timetable shows return times into the valley towns. The afternoon train departed Kingston at 2:45, with arrival at West Hurley at 3:04. The added time is due to the plateau of the Esopus Valley being 350 feet higher than the Kingston Depot. The U&D had a fairly steep climb getting to West Hurley.

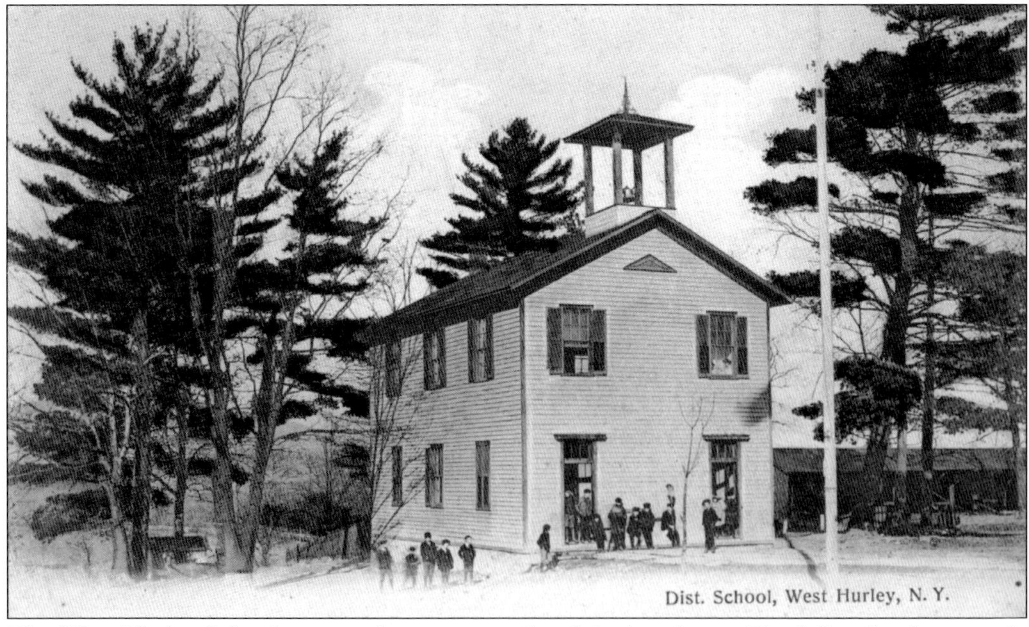

The West Hurley School was on the western side of the hamlet on Plank Road, subsequently renamed Main Street, a few blocks west of Railroad Avenue next to the Methodist church. All grades were taught there. School was in session from September until mid-May. Heating was with a centrally located woodstove, managed by the teacher.

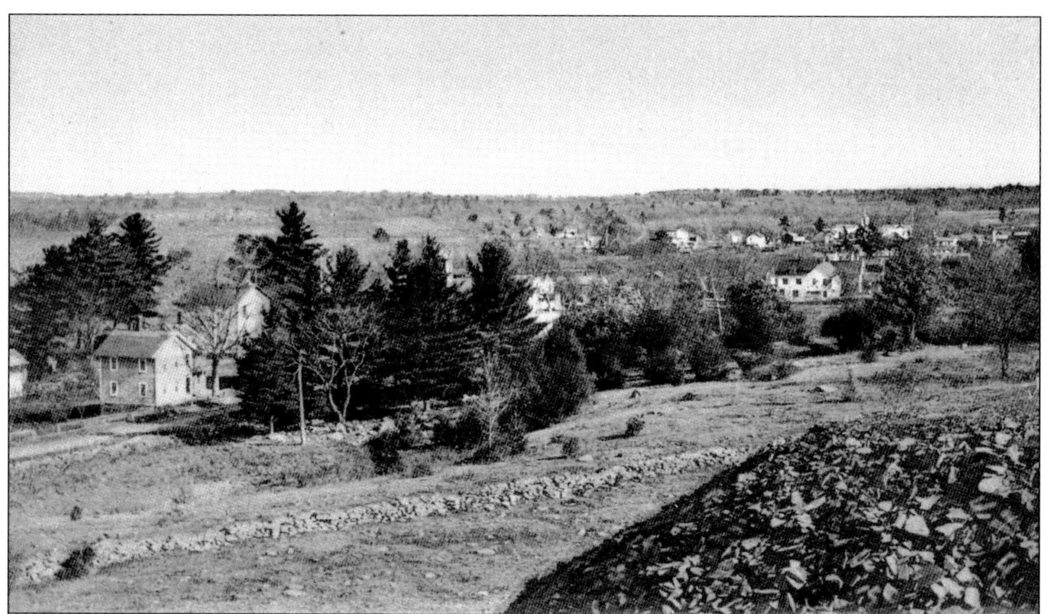

This West Hurley overview taken from the hill west of the depot provides a good view of the village. Most of the buildings are on the right side and make up the center of activity with two churches there. The school was out of the picture to the right.

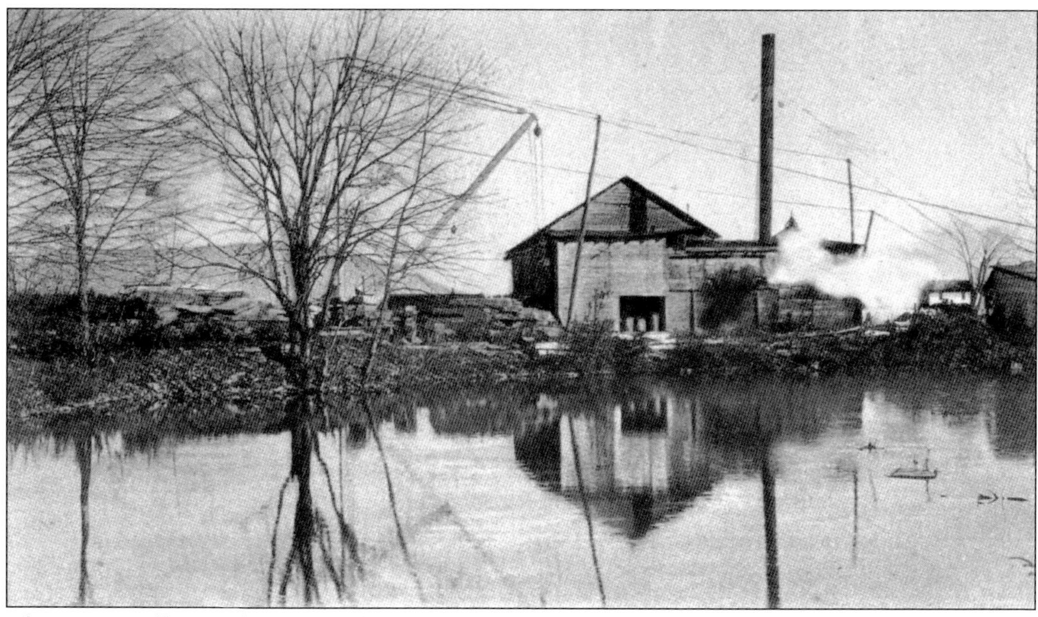

This stone mill was a factory on the outside of the village where quarried bluestone slabs were cut to size. Pieces were then shipped by wagon to Kingston over roads made in the manner shown in the following image. Later, these same quarries provided aggregate used in the concrete mix for the East Basin dikes and spillway.

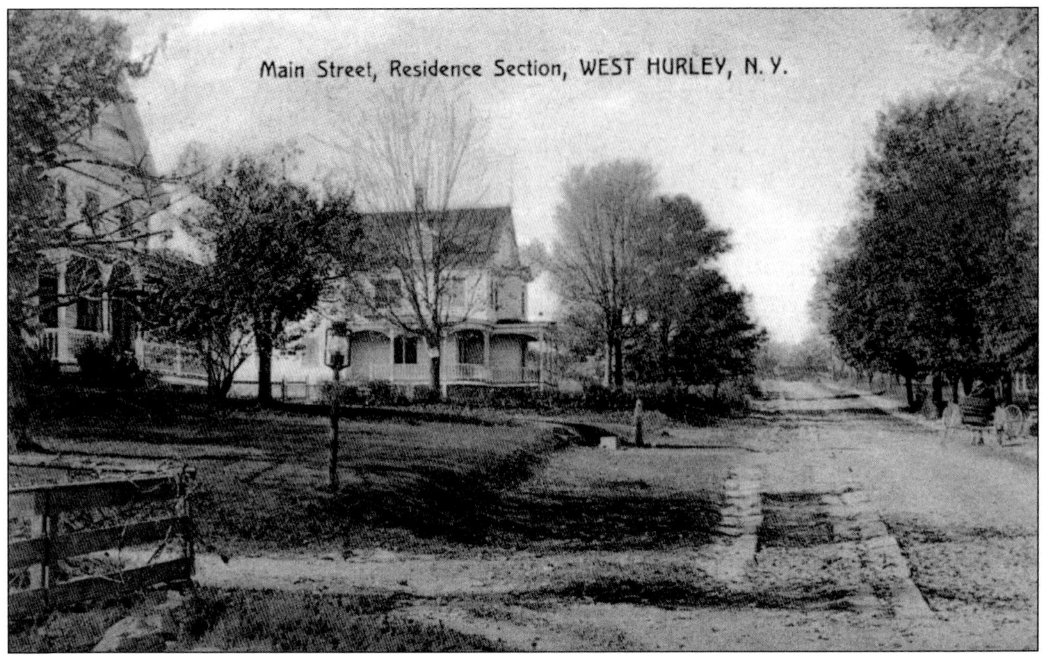

This image of West Hurley shows a bluestone roadbed—thick blocks of bluestone make a surface that allows transport of the stone from the quarry to the work yard in all weather. A close look reveals ruts in the stone caused by the wear of the steel-rimmed wagon wheels. These grooved blocks may still be found in the area. A few are at Frog Alley in Kingston.

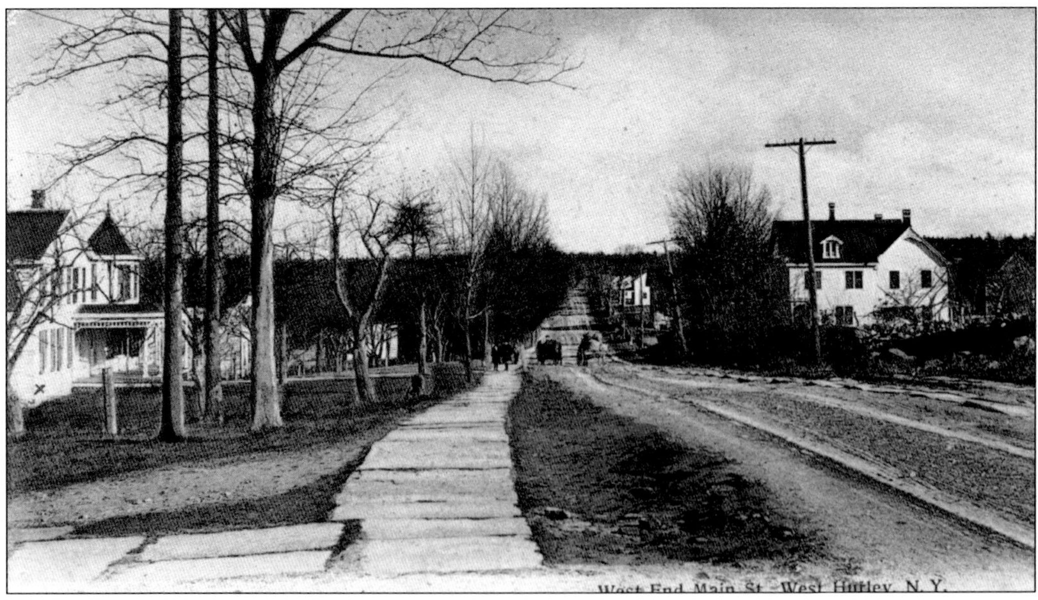

This view of West Hurley looks east from the western end. Most remarkable are the bluestone sidewalks. The rutted road makes it clear why a good walking surface was important. Most local communities still have bluestone sidewalks. According to the US Geological Survey's 1905 maps, a large marsh covered much of the area between West Hurley and Ashton to the west.

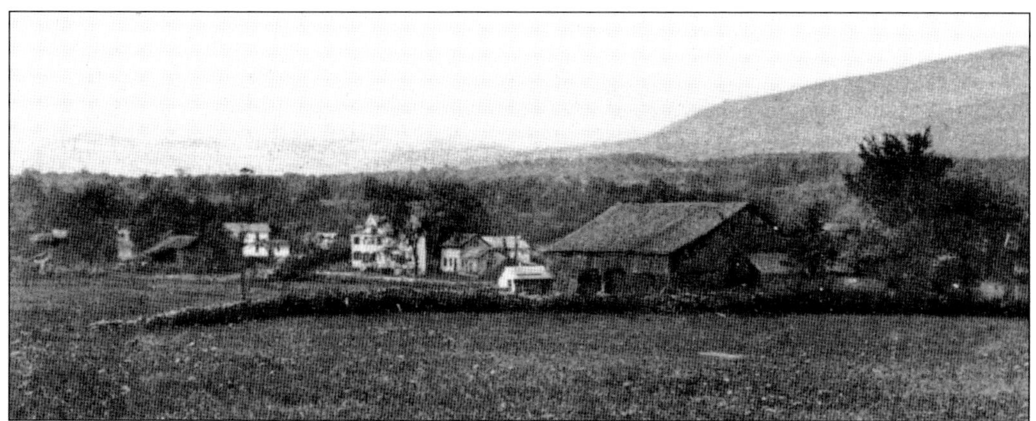

Ashton was a small hamlet with a post office and general store on the north side of the Esopus Valley. Most residents there maintained farms. Some opened their homes to summer boarders from New York City. This view shows the Beaverkill, flowing to the left into the Esopus. *Kill* is the Dutch word for "stream."

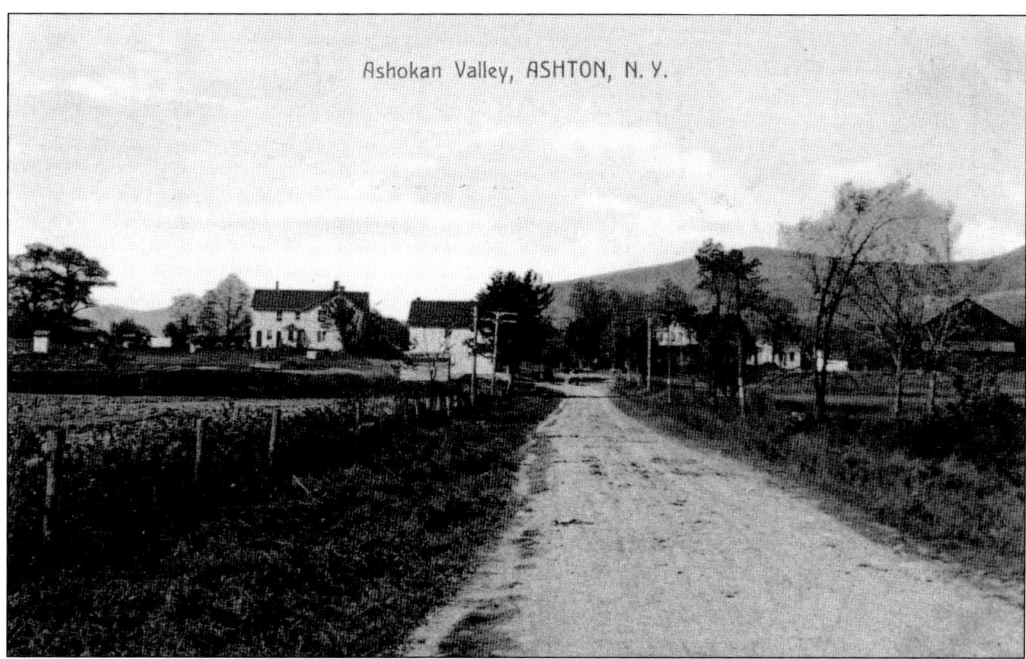

Ashton, in this view looking west, is a small collection of homes, with West Hurley behind and the Olive hamlet ahead. US Geological Survey maps indicate the hamlets were about three miles apart, with very few hills. The U&D crosses the road ahead of this view. About 10,000 to 12,000 years ago, this valley was a glacial lake.

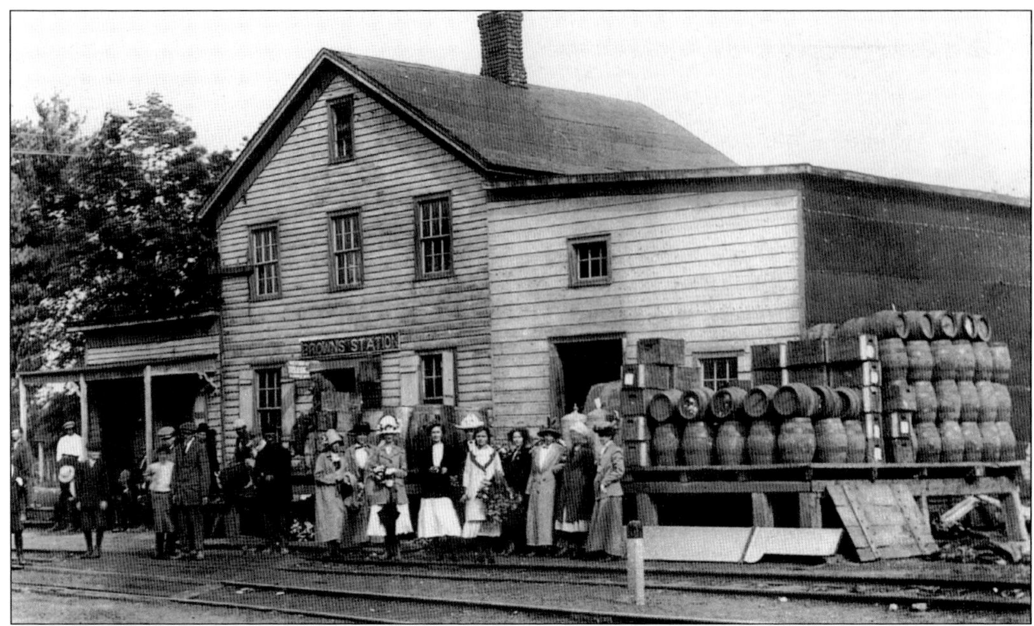

Brown Station was a major rail stop. The village was near the center of this fairly flat valley and was a hub for much of the produce and farm products. Barrel hoops were a major product. This image shows well-dressed travelers and a large amount of produce, probably cider, in barrels, waiting for the train into Kingston. (Courtesy of John Ham.)

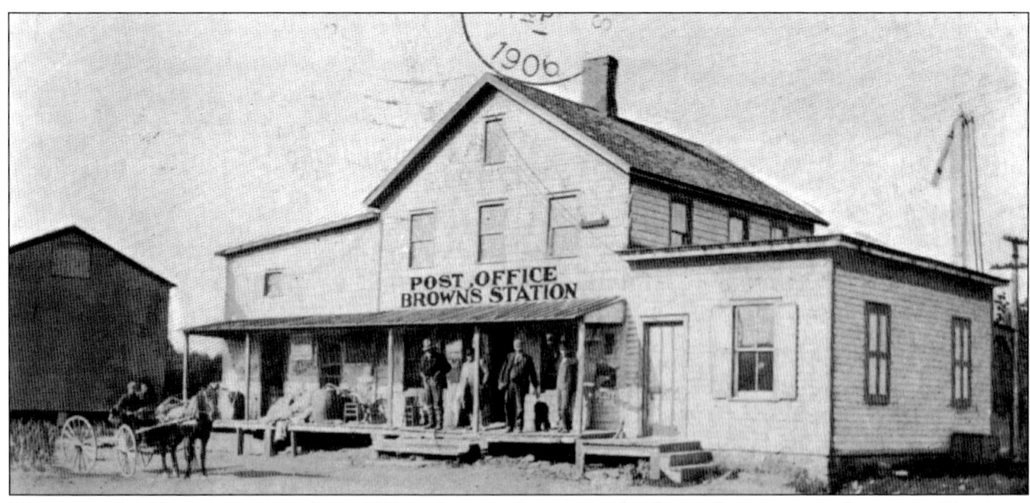

This was the Brown Station Post Office. Located not too far from the station, the post office was a meeting place for town citizens. In many hamlets, the post office was part of the general store or in someone's home, as in Ashton. Brown Station was the center of activity as the reservoir was being built.

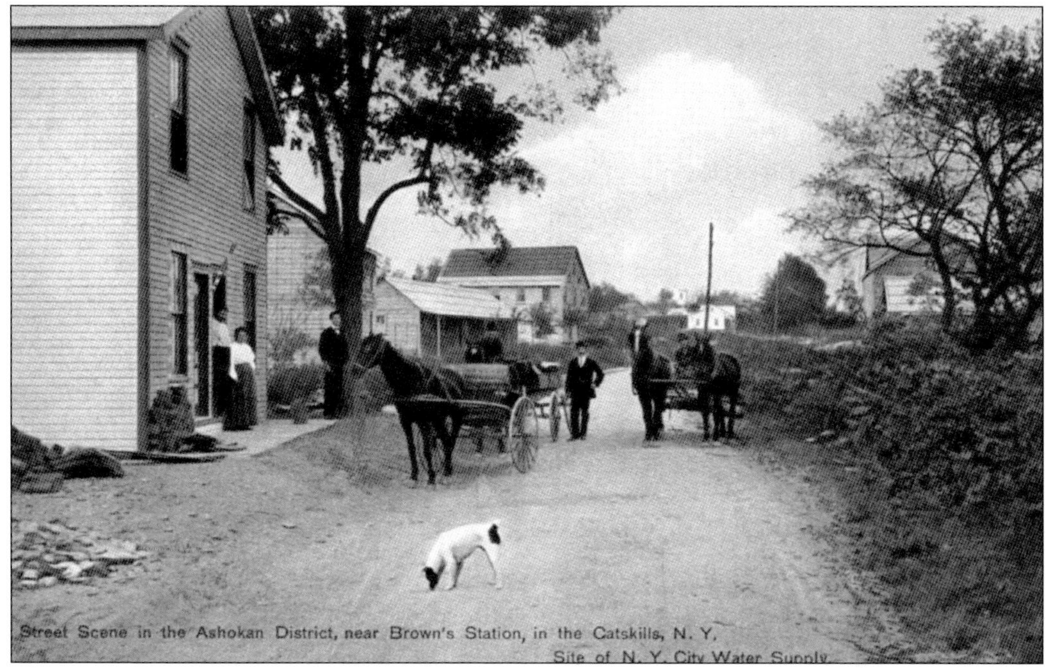

This postcard shows a typical scene around Brown Station. Homes were all two floors with a central wood stove downstairs. Heat for upstairs went through a grating in the downstairs ceiling. Winter temperatures here would often reach well below zero. Summer boarders were welcomed by the hamlet residents.

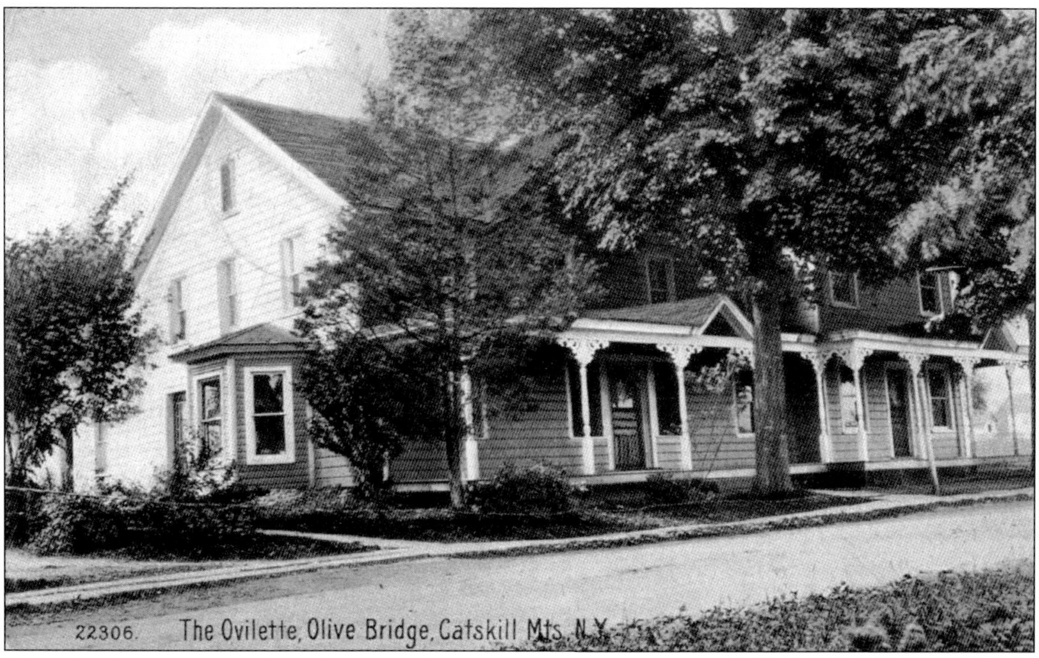

Olive Bridge, on the southern banks of the Esopus near the Bishops Falls bridge and mill, consisted primarily of boardinghouses. Here is the Ovilette, one of many named boardinghouses. There were others, many quite large, on the side roads off the main road.

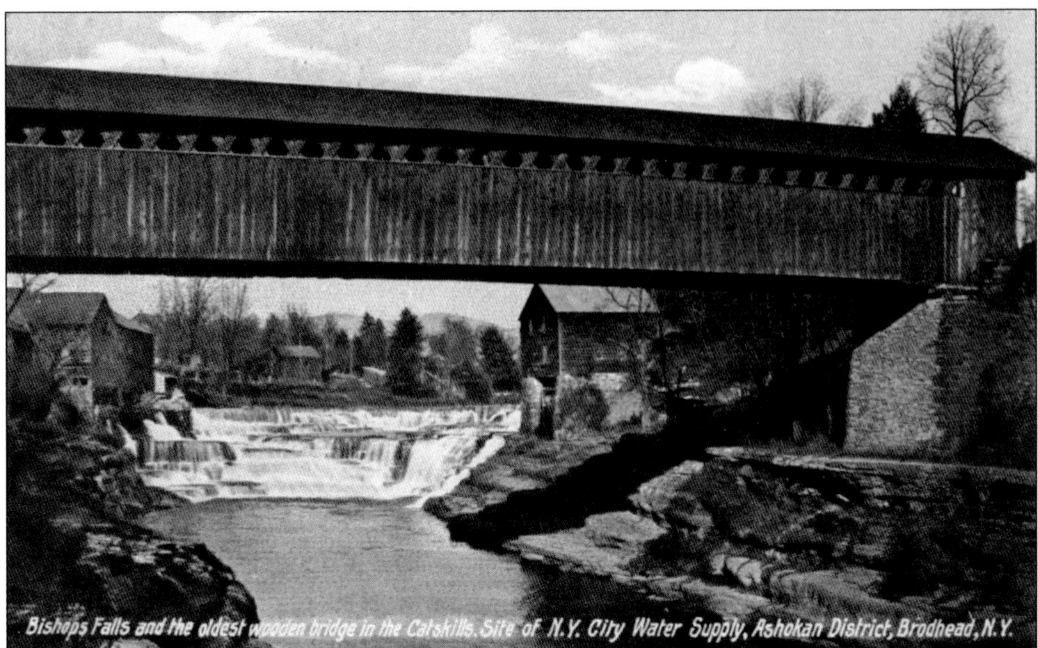

Seen here is one of the few Bishops Falls postcards that included the Esopus covered bridge. The bridge was destroyed in a flood in 1909 during dam construction. This view was one of the reasons New York City residents visited the valley, along with the Esopus gorge, now the location of the dam. This mill, run by a blind miller, and the falls, were the iconic representation of the entire Esopus Valley.

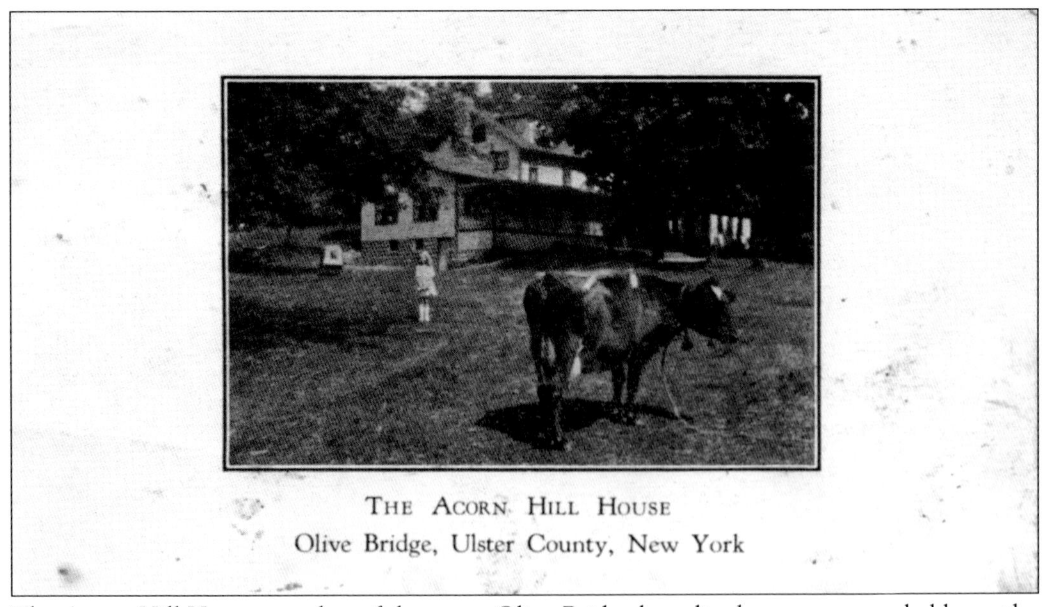

The Acorn Hill House, another of the many Olive Bridge boardinghouses, was probably on the road to Broadheads. The map at the beginning of this chapter shows roads leading away from the creek where most boardinghouses were located. Summer visitors spent their days walking the roads and trails, not in the finery seen at the railroad stations.

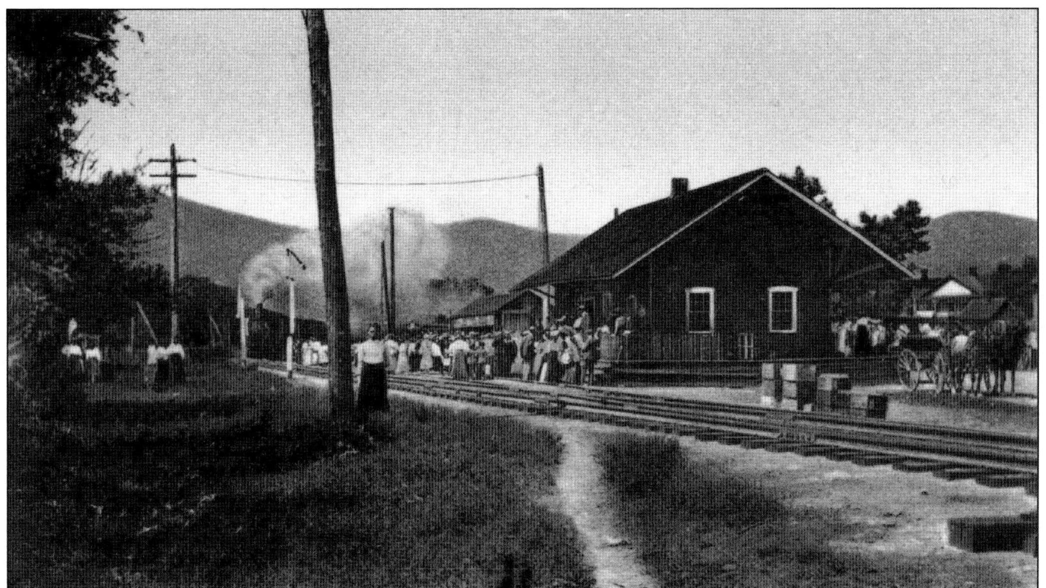

Broadheads station was very busy during the summer months, as can be seen by the number of travelers waiting to board the arriving eastbound train. Notice how well travelers dressed. Trunks of clothing must have followed the summer guests. The trip involved a day boat to Kingston, a train to the station, a wagon to the house, and the reverse at the end of the visit.

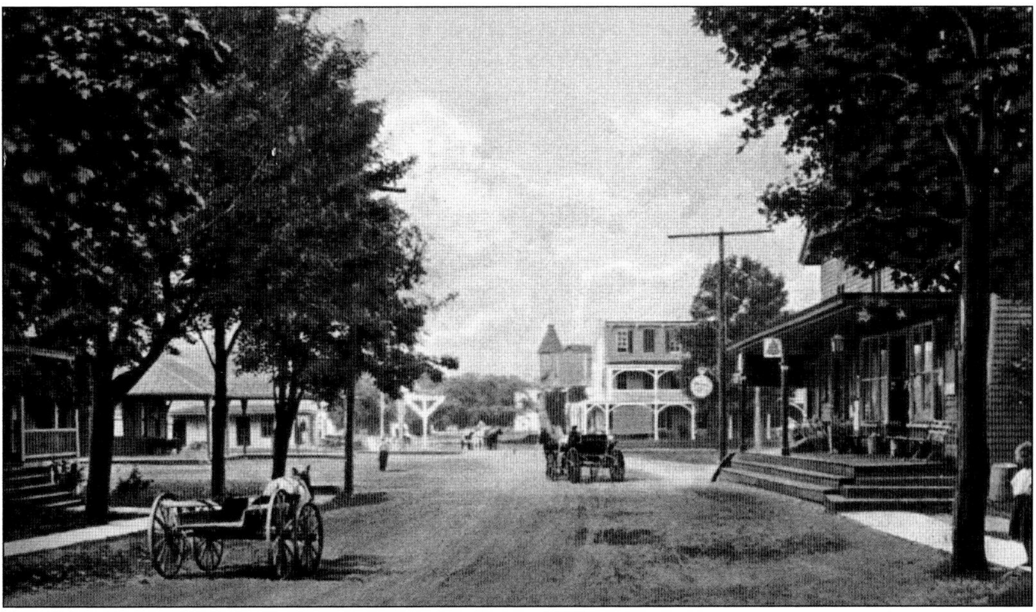

West Shokan is the most western and the largest of the valley communities. The end of the depot canopy is visible through the trees on the left. On the right is a general store. Note the telephone sign on the canopy post. West Shokan was an active community and home of the first eagle scouts in New York State. Five boys earned the highest rank in the Boy Scouts of America on November 1, 1912.

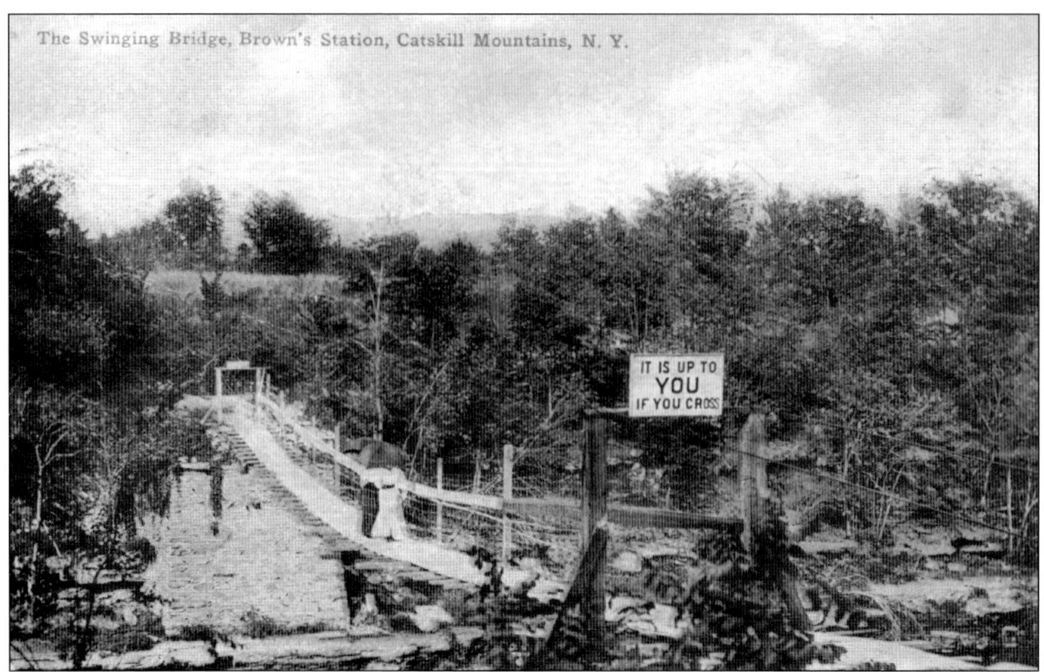

At the head of the valley was the little hamlet of Boiceville. Images of the area are hard to find; however, this view of a footbridge over the Esopus is very popular. As with all suspended bridges, it swayed when being crossed, especially if walking in step or together, hence the sign. Unfortunately it is not there today, as it would be a nice side trip off the rail trail.

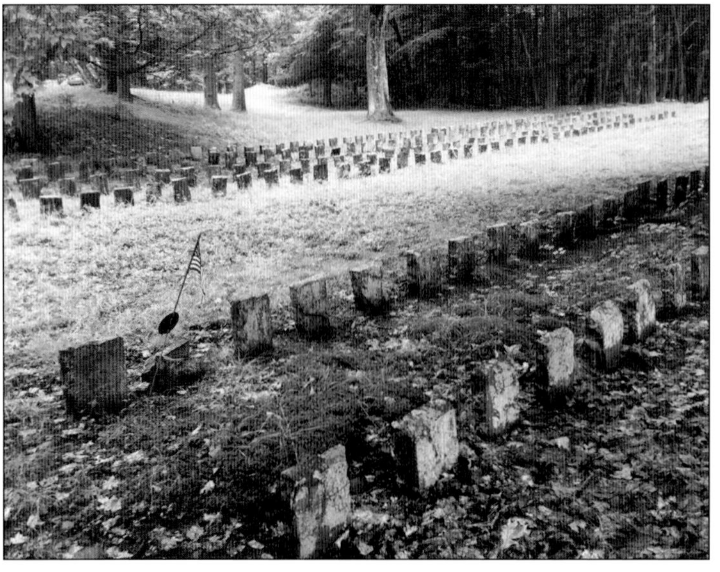

These markers in the Bushkill Cemetery identify each valley resident who had no claimant and provide an indication of the age of the valley. Movers hired by the BWS at $15 each also provided the markers that identify the cemetery and grave location where the body was disinterred. These markers, two churches, and a few nearby buildings moved intact are all that remain of a thriving community. (Courtesy of Kate McGloughlin.)

Two

The Olive Bridge Dam

In 1906, while state government approvals were being sought, exploration and design work began. The valley was perfect for a reservoir, as the bedrock was horizontal. The weight of the dam and water would compress strata, minimizing leakage. Bedrock test borings made during spring and summer helped find the best location. By mid-August, a site a quarter-mile below Bishops Falls was selected for the waterside of the dam. The rate of water flow in Esopus Creek made it necessary to design a means to contain it during construction. The BWS contractors established a cofferdam and a stream control method while detailed dam designs were being put to paper. Construction contracts were awarded on August 31, 1907, with work starting on September 30.

Winston & Co. joined forces with McArthur Brothers. The pair recently completed the Kensico Dam, the terminus of the Catskill Aqueduct. A powerhouse, cement plant, and large stone crusher had to be moved to Olive Bridge and assembled. A workers' camp was necessary before men could be brought to the site. Four cableway systems, similar to those used on the Panama Canal, were installed across the dam site. Both narrow-gauge and standard-gauge work trains were added to the site. A three-mile standard-gauge rail line ran from the cement plant to the Yale Quarry. The quarry, worked for years by valley residents, provided crushed stone and paving stone until the project was completed.

MacArthur Brothers and Winston & Co. had a higher bid than others; however, with their experience building the Kensico Reservoir, they were awarded the $12.7 million contract to build the dam. Much city hall wrangling went on, but the decision stood. In September 1907, work began. The dividing weir and adjoining dikes were included in this contract. Weidner writes in *Water for a City* that at the end of 1908, the contractor population was 2,455 men, 178 women, and 306 children.

This chapter will discuss the construction of the dam and both wings. The eastern dikes, perimeter roads and bridges, and land clearing were separate contracts.

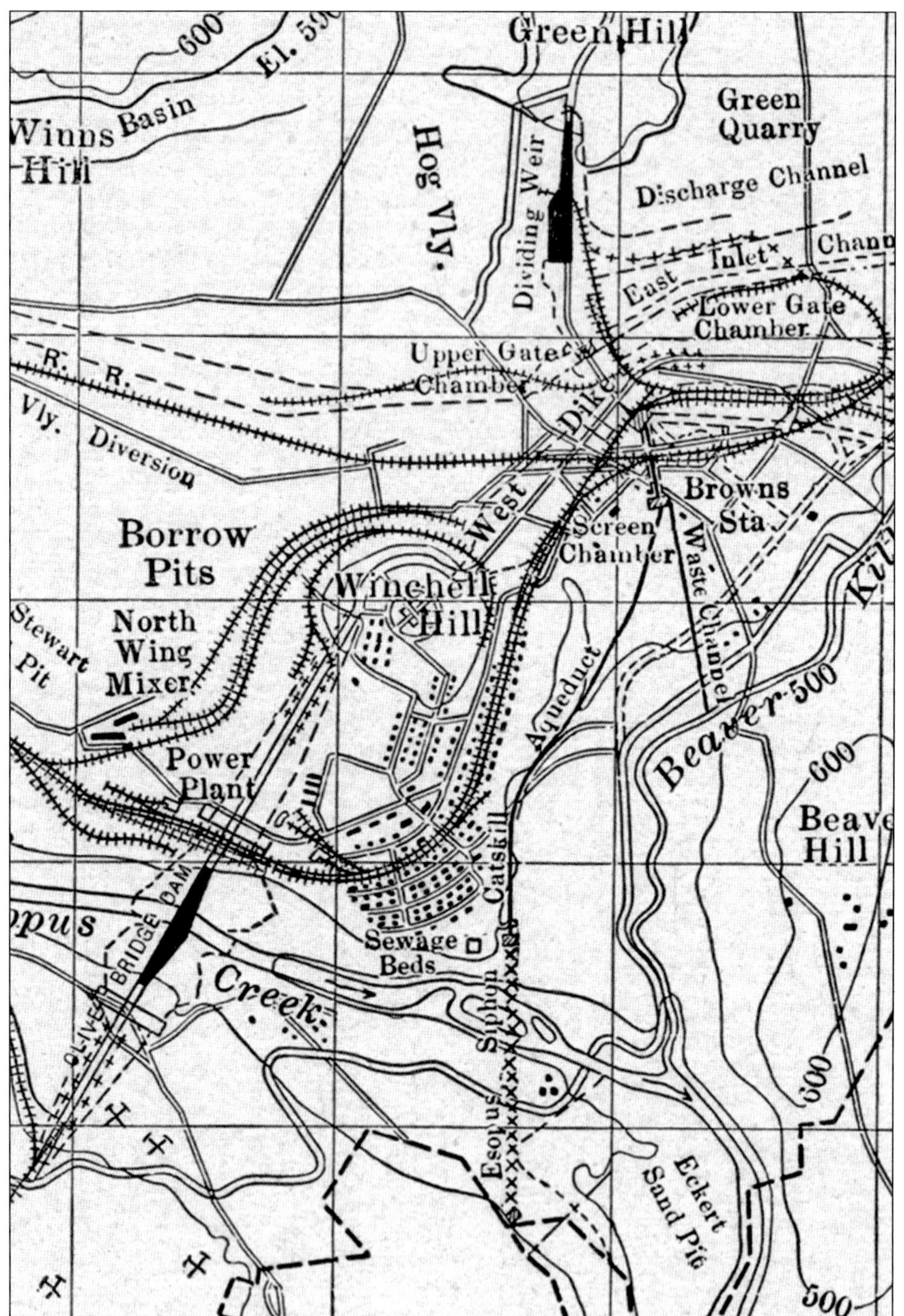

This reservoir works map is from Lazarus White's *Catskill Water Supply of New York City*. The Main Dam, a black line crossing the Esopus, is at lower left. The dashed lines at both ends are wings with embankments. The workers' camp is below Winchell Hill. The power plant is to the left. The wide black line at top center is the dividing weir. A multitude of railroad tracks and roads are shown. The U&D Railroad crosses below the dividing weir.

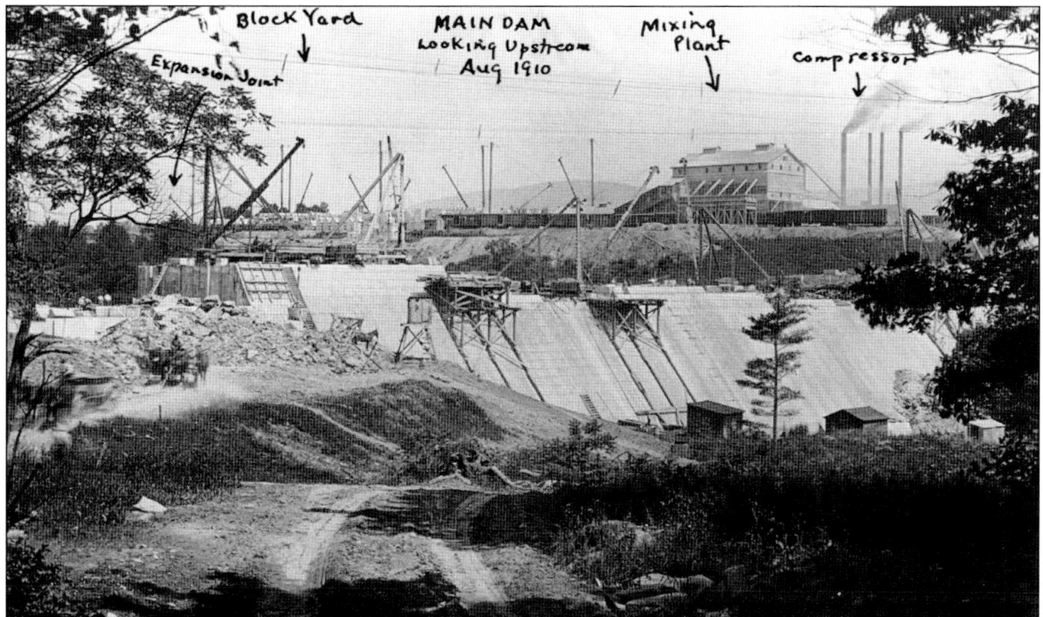

This postcard identifies the key parts of the site and shows the dam nearing completion. The block yard, at far left, stored manufactured blocks for three months prior to use. Blocks were brought to the dam by an overhead cableway out of view on the right. Similar blocks found their way into today's bridge and retaining wall construction.

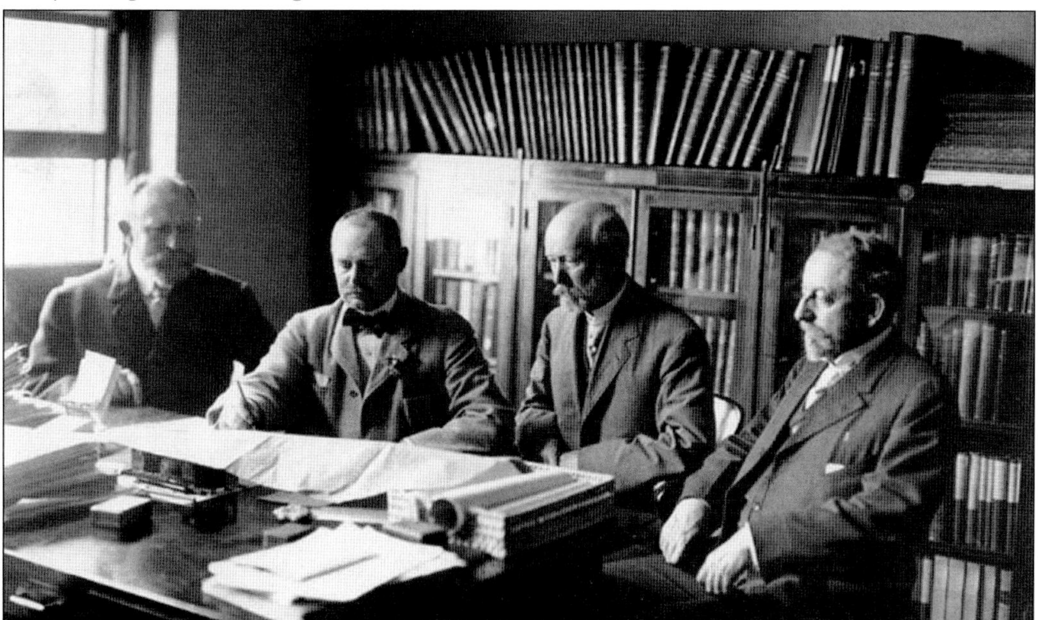

The lead project engineers were, from left to right, J.R. Freeman, a BWS consulting engineer for many years; J. Waldo Smith, chief engineer; F.P. Stearns, consulting engineer; and Prof. W.H. Burr, who graduated from Rensselaer Polytechnic Institute in 1872, then a civil engineering professor at Columbia. Smith worked on water supply systems at 15 and graduated from the Massachusetts Institute of Technology in 1887. He joined the BWS aqueduct commission as chief engineer in 1903 after working on various water construction projects in New York and New Jersey.

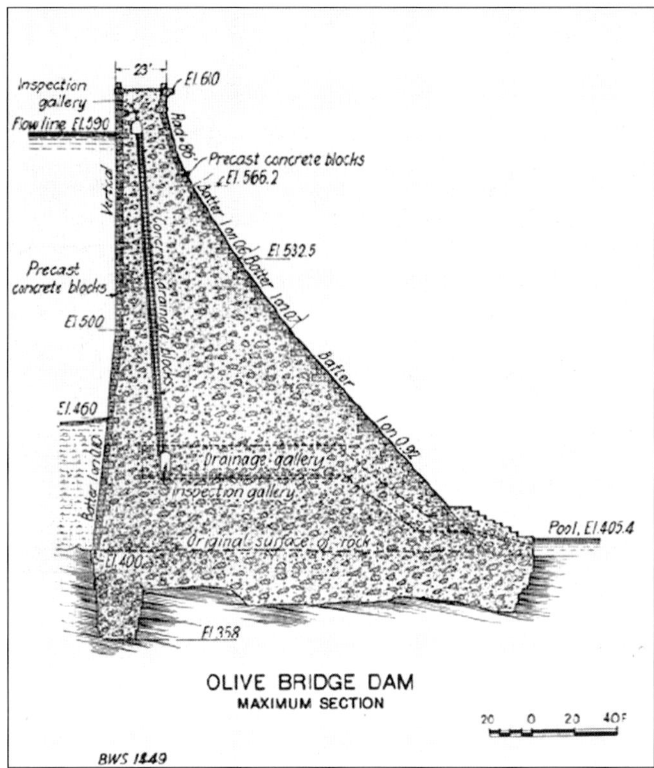

This cross-section shows construction methods and elevations in feet above sea level. The finished height is at an elevation of 610 feet. The water level at full pool is 590 feet. The foundation started at 400, not including the cutoff (lower left). The surfaces are precast blocks, eliminating the need for forms. The structure has inspection galleries and drainage channels.

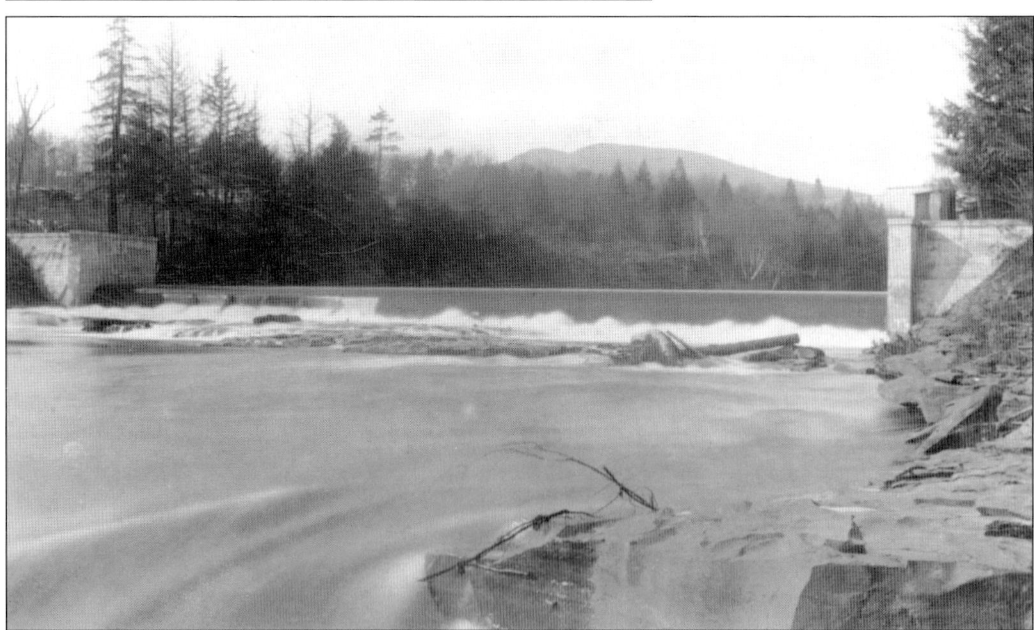

A measuring weir is used to determine the volume of water flow over time. The width and stream flow rate were known, and the depth of water over the weir was measured. The water flow was calculated in cubic feet per minute—the product of flow rate (feet per second) and flow area (width times depth). The diversion pipe diameter was based on the highest flow rates. Two eight-foot-diameter pipes were deemed adequate.

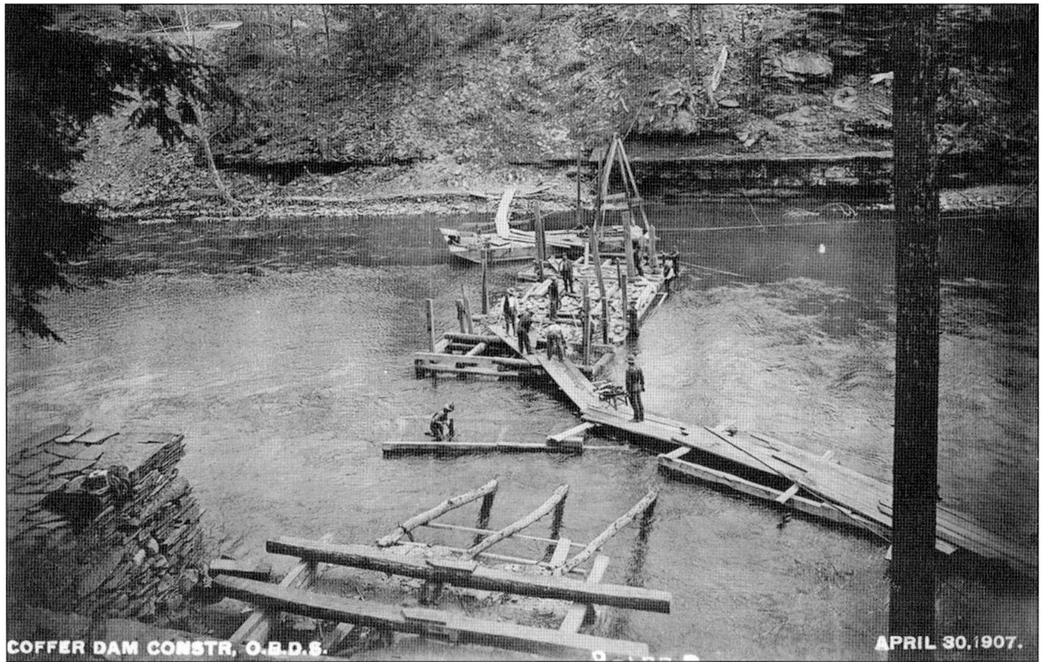

Before construction contracts were approved, BWS built this cofferdam to force water to flow to one side of the creek. Containers full of stone were dropped to the stream bottom, and filling the width to the north shore completed the dam. This photograph was taken on April 30, 1907. (Courtesy of NYPL digital library.)

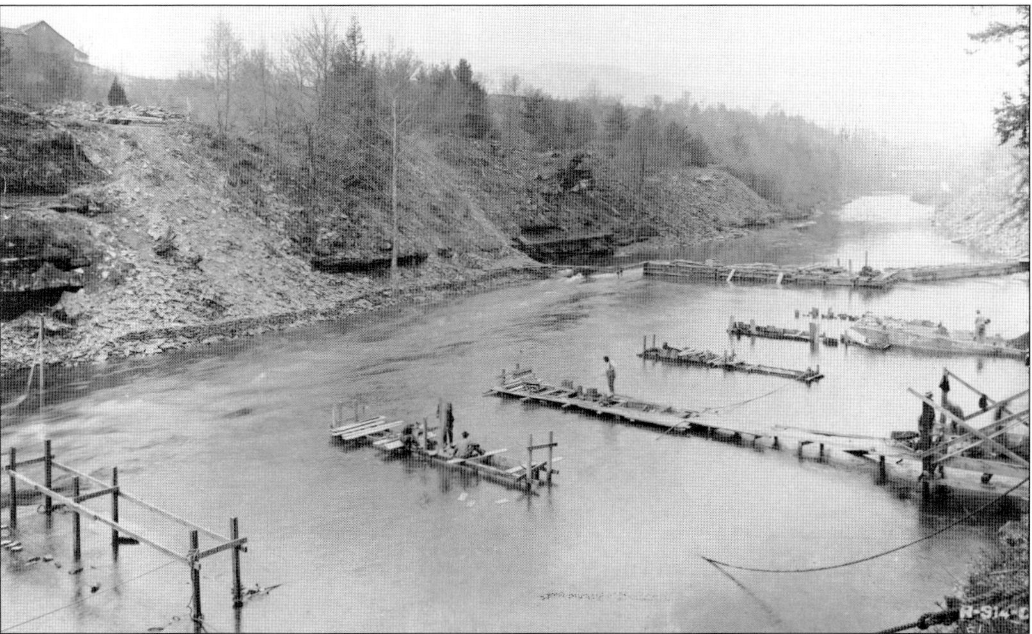

The cofferdam diverted water to flow along the south bank. The resulting quiet pool allowed workers to build supports for the diversion pipes. Wooden framing made "boxes" that were filled with stone and concrete to make solid bases. This photograph was taken on May 18, 1907. (Courtesy of NYPL digital library.)

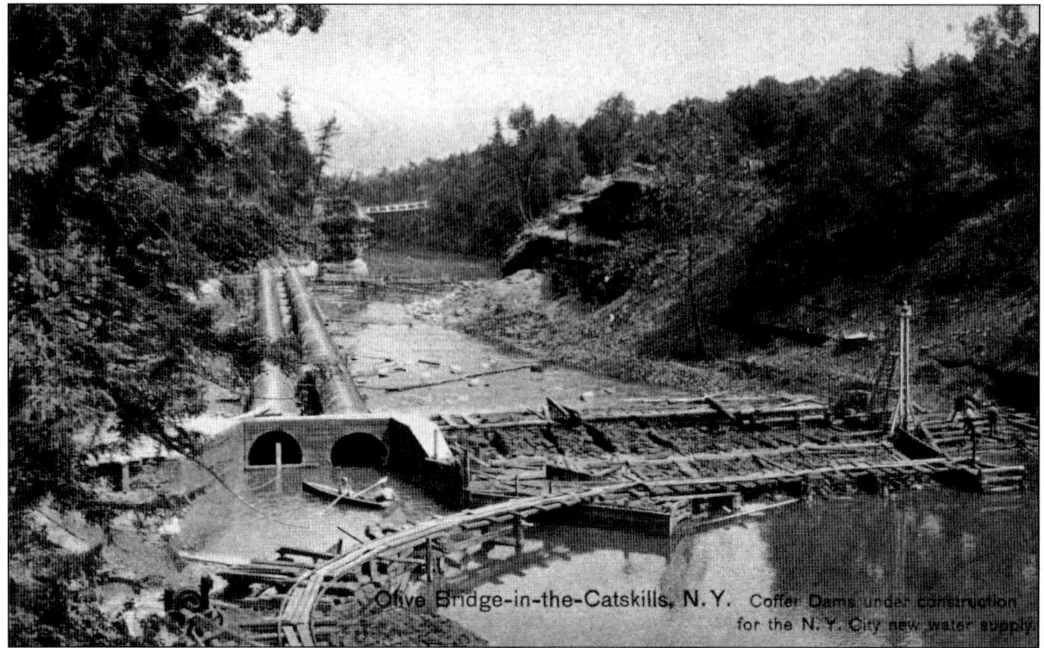

This is a very popular postcard. The diversion dam and pipes, 140 feet long and 8 feet in diameter, are shown in place. The pipe bottoms are at 410 feet elevation. Narrow-gauge trains were used to bring fill to block the water flow. Pumps on the downstream end were used to remove water from the work area. A surveyor's tower is on the right. This area became the dam foundation.

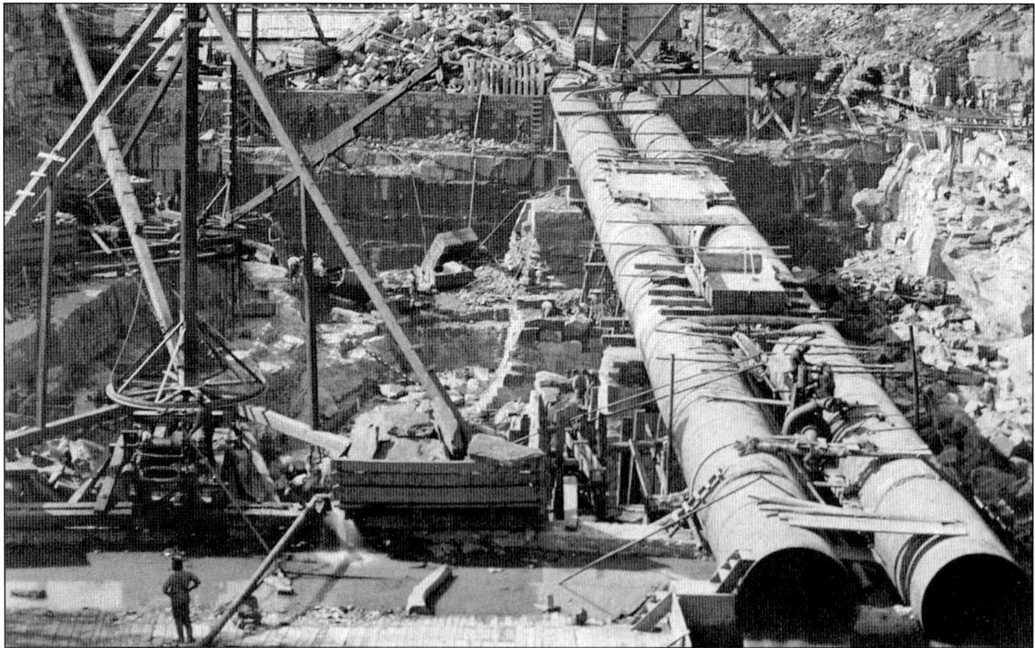

Here is another popular postcard, not as well understood. Derricks are lifting quarried bedrock out of the stream bed for removal by overhead cableway. Compressed air hammers rather than explosives were used to prevent cracking the remaining stone. All loose stone and debris had to be removed before the foundation masonry began. The entire project used compressed air tools at 80 psi.

This image shows the nearly finished cutoff trench, with a 20-foot bottom width and 40 feet where the full-width dam foundation begins. Steps in the side walls and grooves in the end walls allowed masonry to lock into the base rock as the trench was filled. Bedrock cracks found this deep were drilled an additional 30 feet and grouted. The cableway towers are visible on the ridge behind the cutoff.

This view of the dam shows the cutoff. The completed wall of cyclopean masonry under the dam was designed to prevent leakage. With the full reservoir at 590 feet elevation, approximately 11,250 pounds of pressure is exerted at the heel of the dam. Cyclopean masonry fill allows for continuous building without weak concrete joints. Boulders in the concrete mix act as locks between the layers.

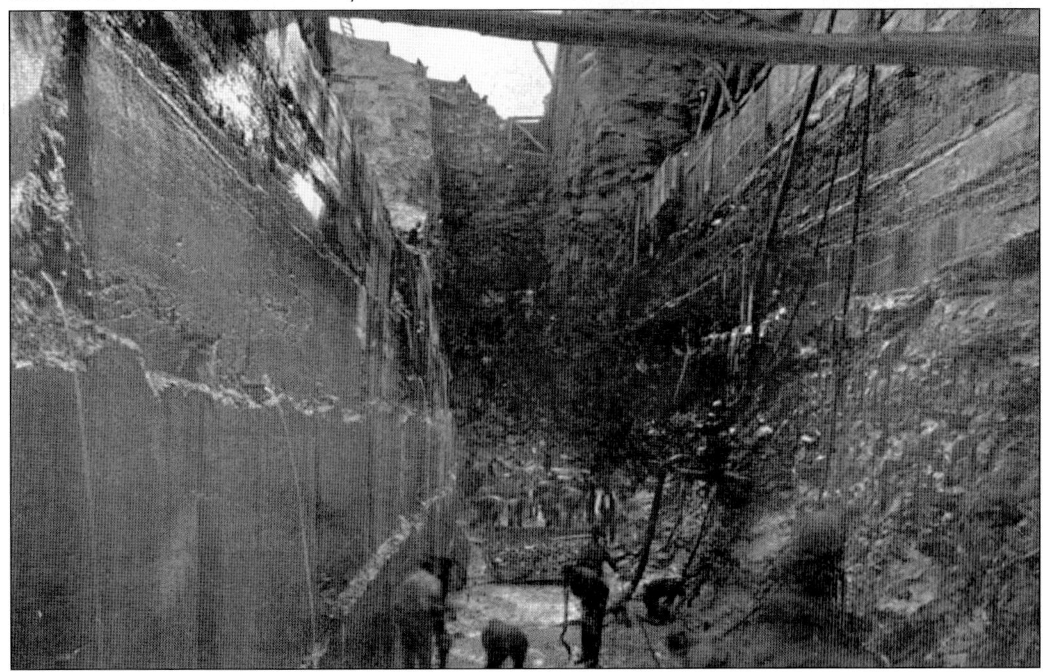

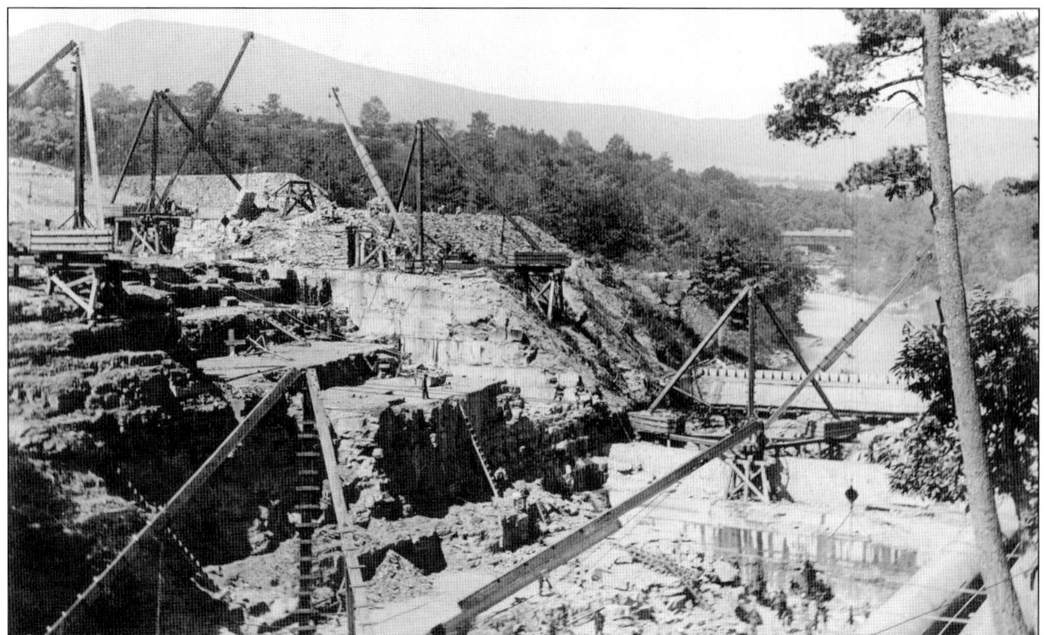

While the cutoff was being completed, a 1,000-by-200-foot area was excavated to solid rock for the dam. Workers shown here are cutting back the walls of the Esopus canyon, similar to the exposed layers upstream of the dam. A chart of the amount of rock removed can be seen on page 35. A quarter mile upstream is the covered bridge near Bishops Falls.

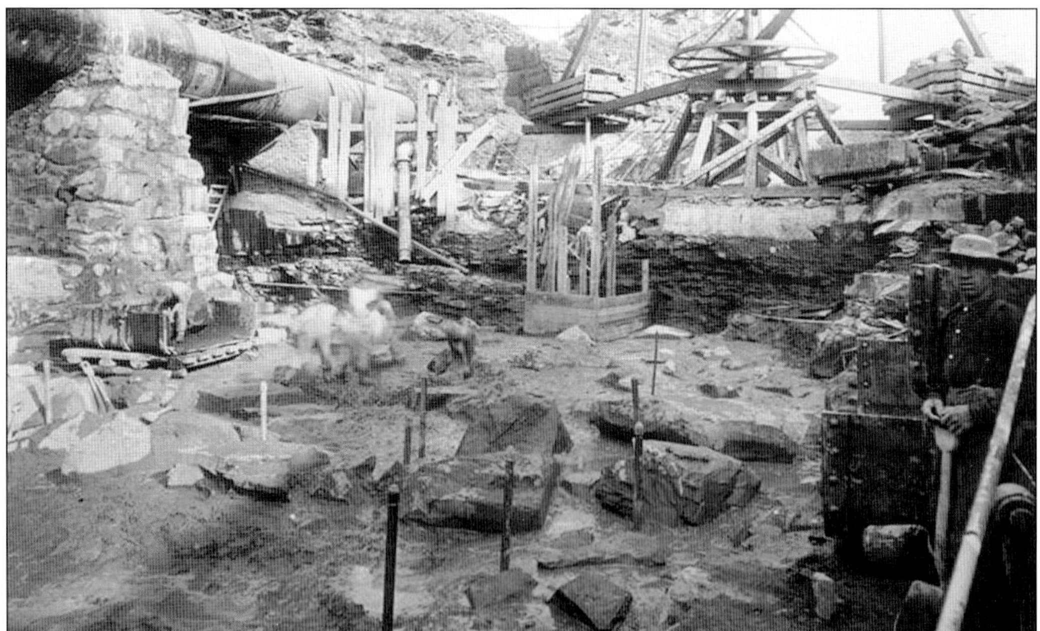

In this image, the trench excavation is completed, and the base area has been cleaned of all loose material. Rock walls made forms for cyclopean masonry. Large grout-coated stone was cleaned and placed into a load of dry mix concrete. The stone was joggled to settle into place, and markers were placed. Another stone was placed a short distance away and handled in the same manner until the foundation reached 410 feet elevation. This method allowed for faster masonry construction.

Here, workers set the first of thousands of precast blocks in the front of the finished foundation. The foundation would soon be built up to where the pipes used for stream control could be removed, allowing water to flow through the dam in the designed method. Cyclopean masonry can be seen behind the front workers. This image may have been captured on October 11, 1909. (Courtesy of NYPL digital library.)

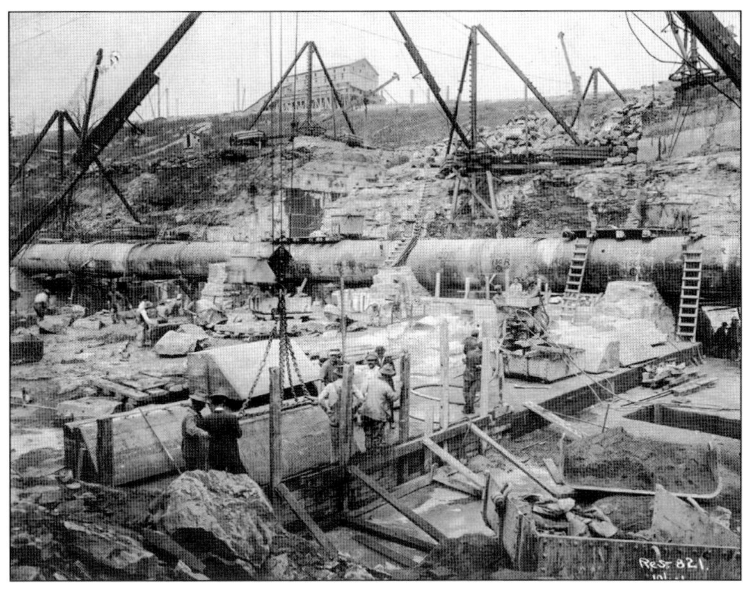

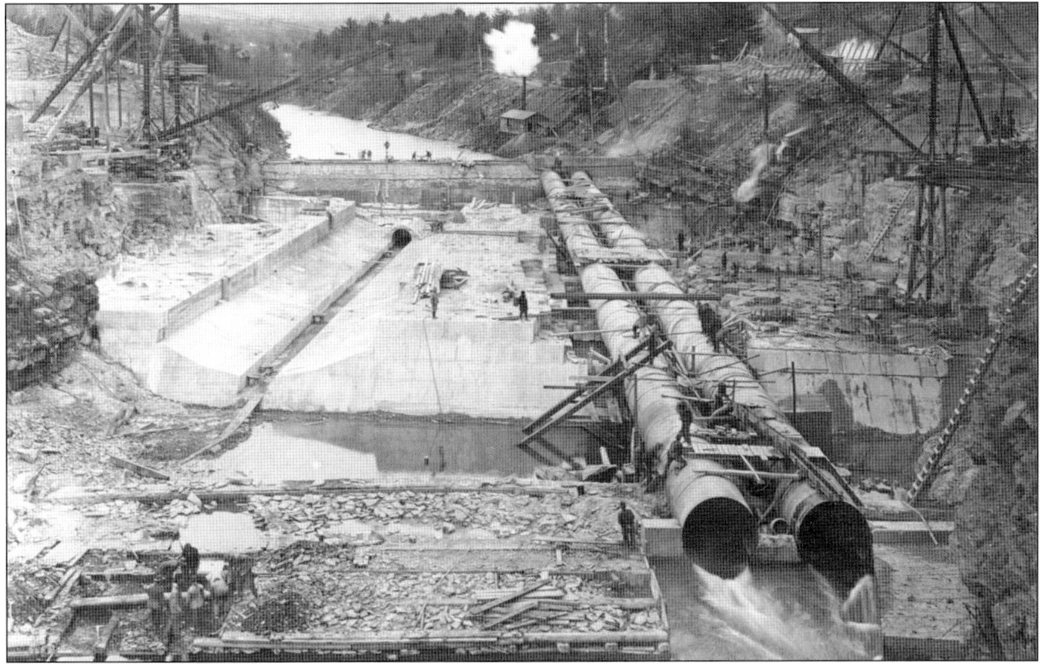

The foundation has been completed to the bottom of the flow control conduit that runs the width of the dam. The 35-foot-wide concrete "invert" is adequate to carry water when the pipes are removed. Notice the six-foot sluice in the center that would be used when the dam was sealed. The bottom of the sluice is at 403 feet elevation. The dam is about 120 feet wide at this point.

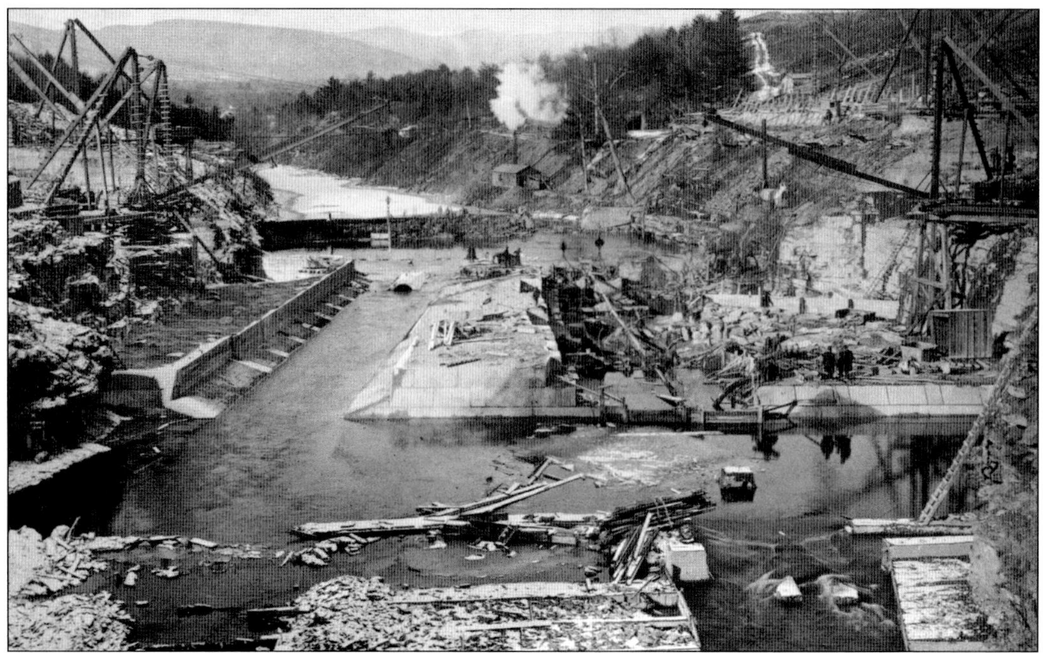

The upstream cofferdam is opened. A rectangular hole can be seen about midway on the dam. Water is now flowing across the conduit invert. The first two rows of set precast blocks laid in late September are visible on the front of the finished invert. Additional height was added to the left side of the conduit. This photograph was taken on December 5, 1908. (Courtesy of NYPL digital library.)

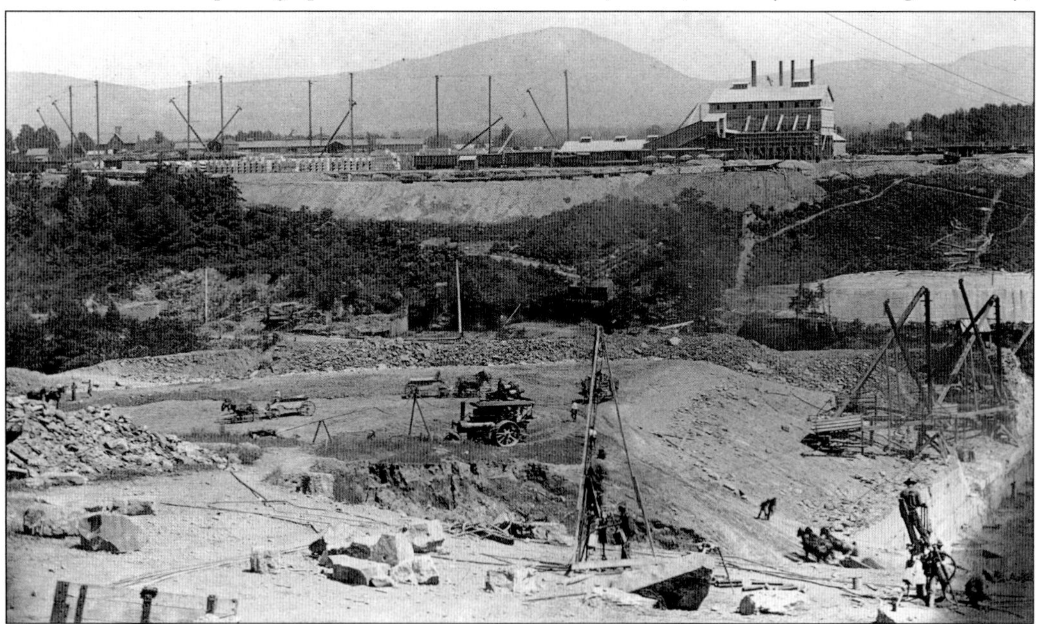

While foundation work is underway, this image shows excavation for the south wing dike core wall and masonry dam. Soils are sorted for embankment use as the dam grows taller, and surveyors are setting lines for rock excavation limits. Behind the surveyors, mule wagons are moving overburden soil. A steam-powered bulldozer levels the material. The concrete mixing plant and block storage are on top of the ridge.

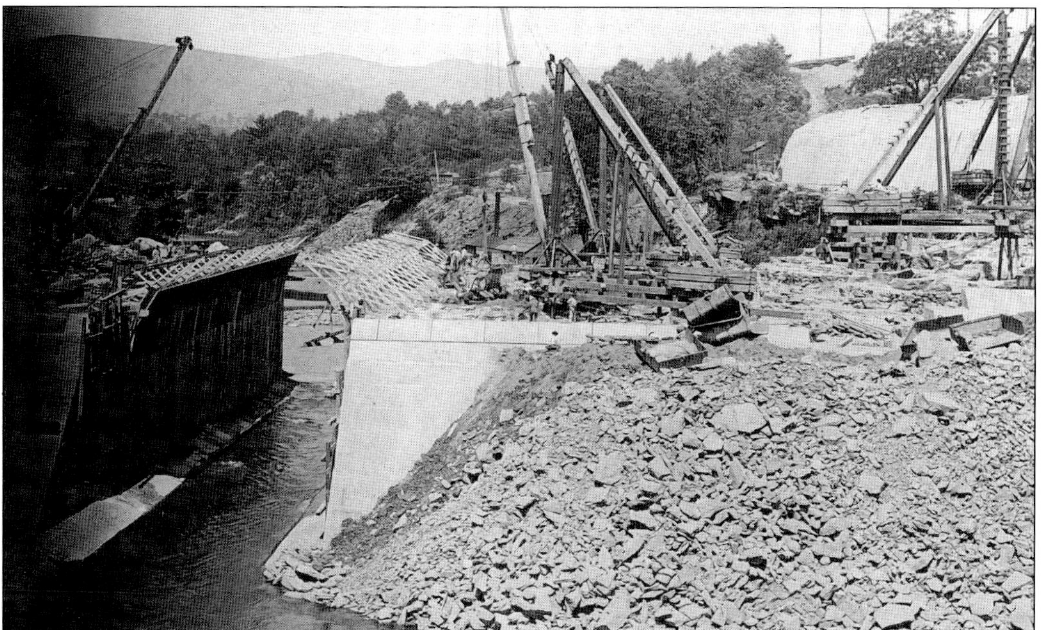

As the conduit reached the proper height for the archtop, wood framing was added to support the masonry. Internal forms could not be used due to the possibility of floods washing away the supports. Concrete delivered by cableway was used to encase the supporting steel. Enough regular concrete was used to make a smooth base for the continuation of cyclopean masonry. (Courtesy of NYPL digital library.)

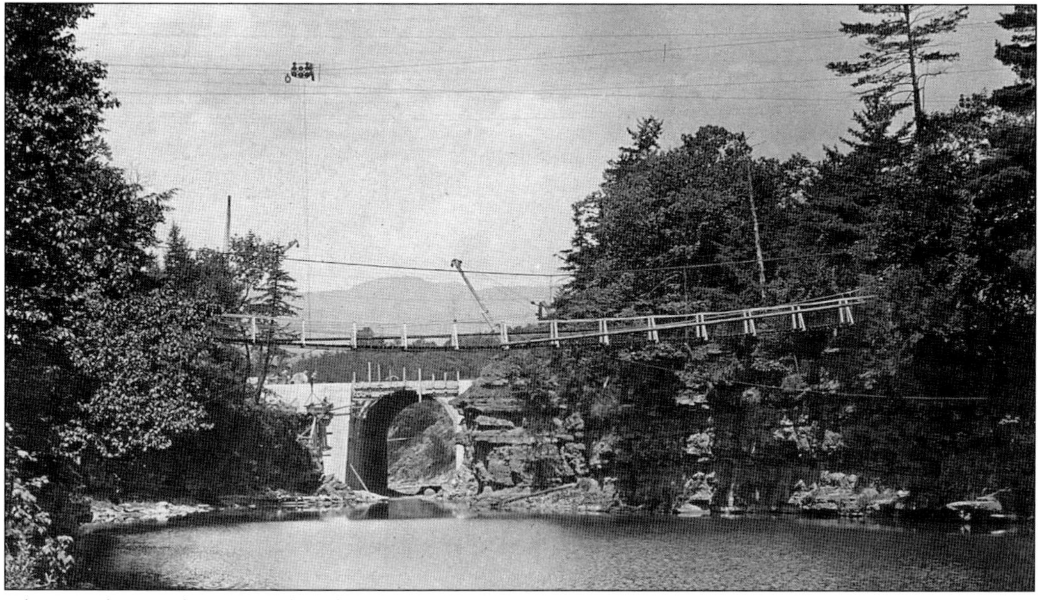

The conduit arch is completed with another layer of block being laid. The workers' footbridge across the Esopus is in the foreground, and a cableway traveler is above. Lazarus White states that cement for all work came by rail in bags from the nearby Alsen plant and was unloaded by hand and stored in an 18,000-bag storage house. Workers emptied the bags directly into mixer hoppers.

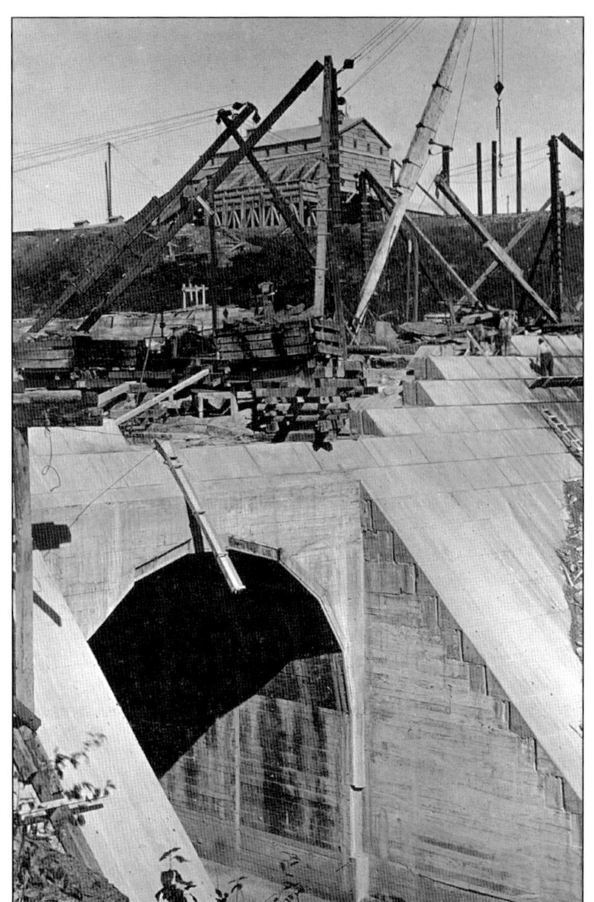

This view shows the completed conduit, which will be finished to match the dam face after closure. On the right, a worker applies mortar to the gap between the top of the lower block and the newly placed face block. Vertical block channels are filled with grout. The July 11, 1911, Red Hook, New York, newspaper stated that in June, 42,000 barrels of cement were shipped to the site from the Alsen plant.

This overview of the dam shows its height slightly above the Esopus Creek canyon walls. An additional support structure was added to distribute the load as the dam above the conduit area grew. Sub-soil that was removed to reach bedrock was stockpiled for later embankments.

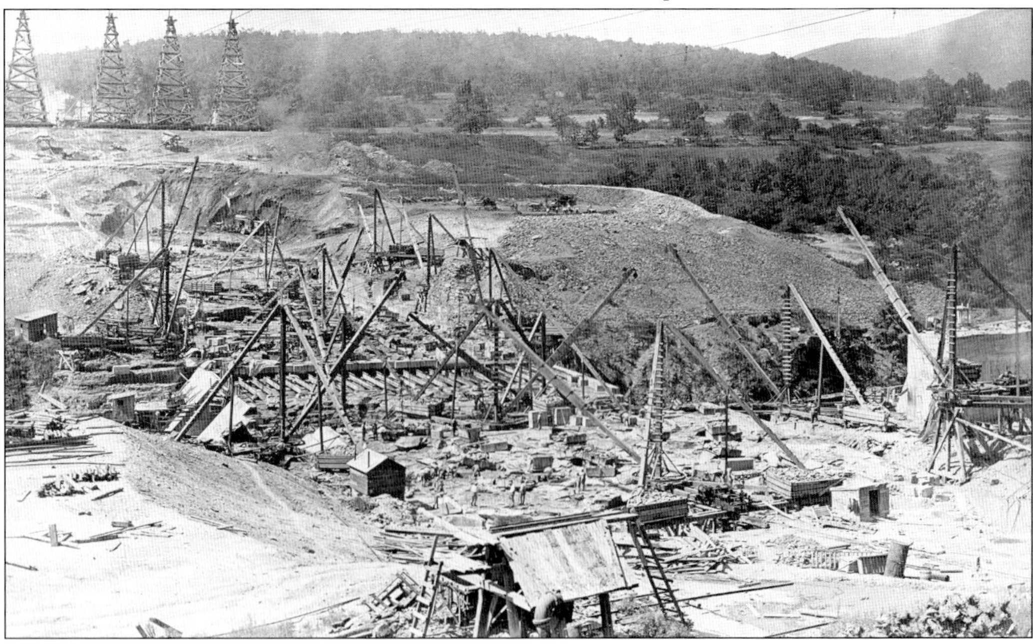

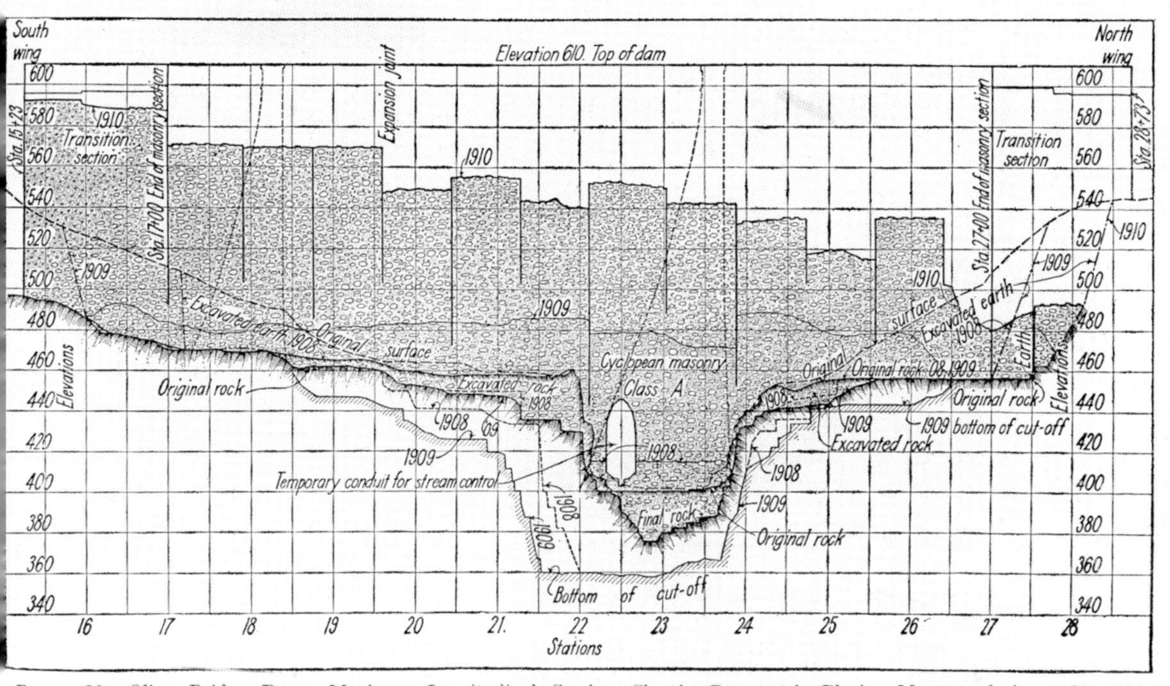

PLATE 30.—Olive Bridge Dam. Maximum Longitudinal Section, Showing Progress in Placing Masonry during 1908, 1909, and 1910. Unshaded Portion was Finished in 1911.

THE ASHOKAN DAMS AND RESERVOIRS 123

This progress report chart shows the work status on excavation and masonry build. It also shows the longitudinal plan of the dam masonry, including the 12 construction sections. Being built in sections, the precast blocks function as forms. During the long daylight of summer, the contractor ran three eight-hour work shifts, with two doing construction and the overnight shift stockpiling materials.

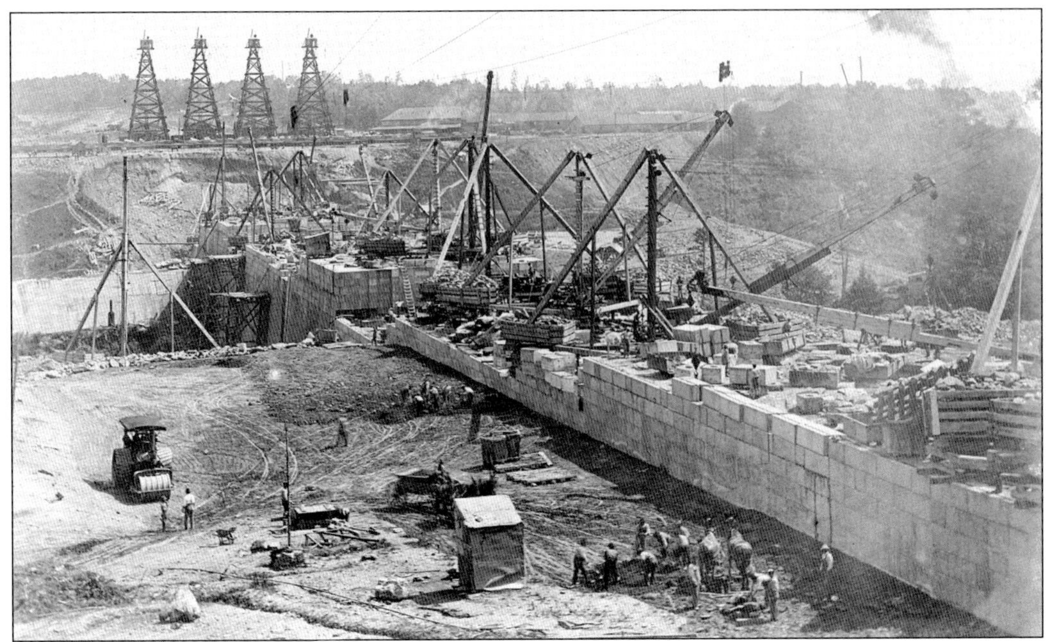

The overnight shift used the cableway to resupply precast blocks. Workers placed blocks using derricks. Cyclopean masonry filled the interior of the dam body. Blocks used for the transverse wall sections were mortared together to provide keying as masonry was added, forming a solid structure (see drawing on page 38). Crews leveled the protective fill against the completed wall, where it was compacted to specifications.

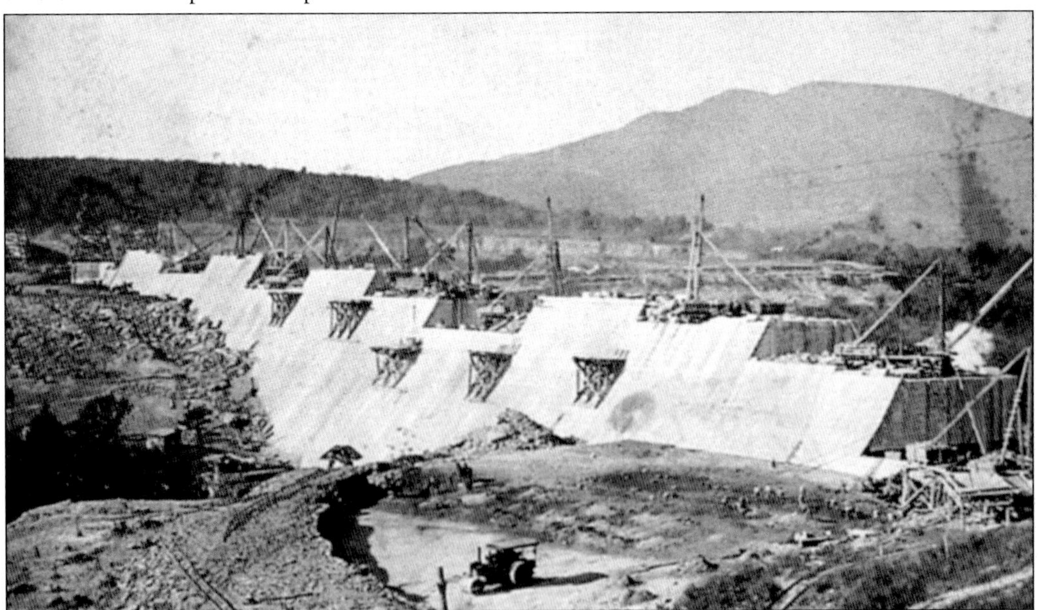

This view shows the dam growing by design in 12 individual sections. Precast blocks make three sides, starting with the first at the south end (left). The opposite row in each section is made so that when stacked, it allows for a vertical opening in the dam body. This channel permits surface water and joint seepage to drain to the lower inspection gallery at 430 feet elevation that extends between the dam sections.

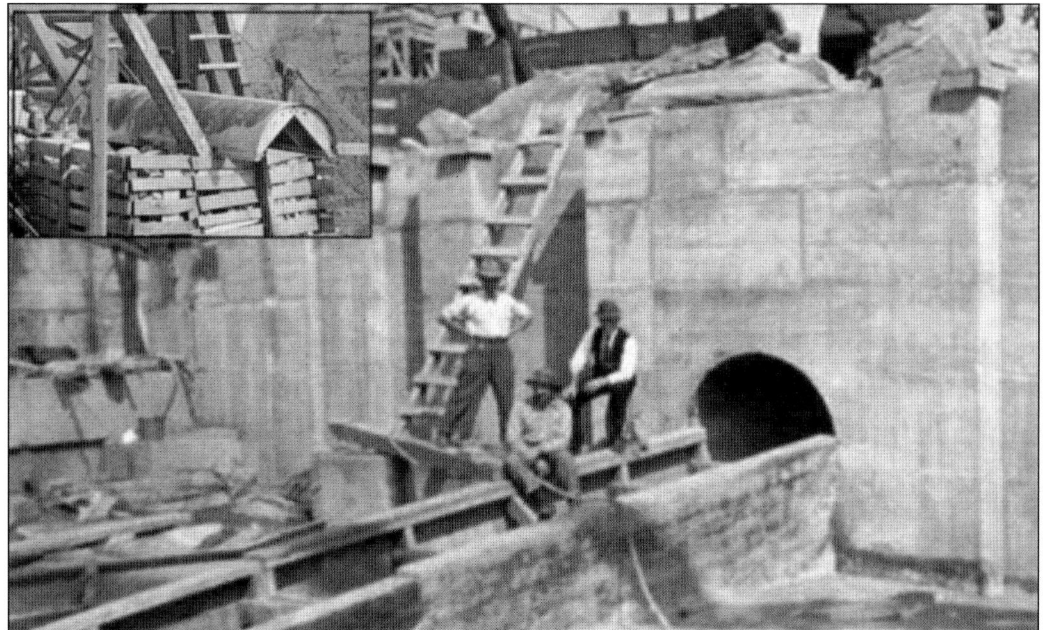

This view shows an opening into the upper inspection gallery. This six-foot-wide, 1,000-foot-long tunnel connects the first dam section with the last. The men shown, a BWS engineer wearing a vest, a supervisor, and a worker, are waiting for the upper curve form (inset) for continuation of the gallery. The wall shape is part of the 11 expansion joints and includes surface drainage into the lower gallery.

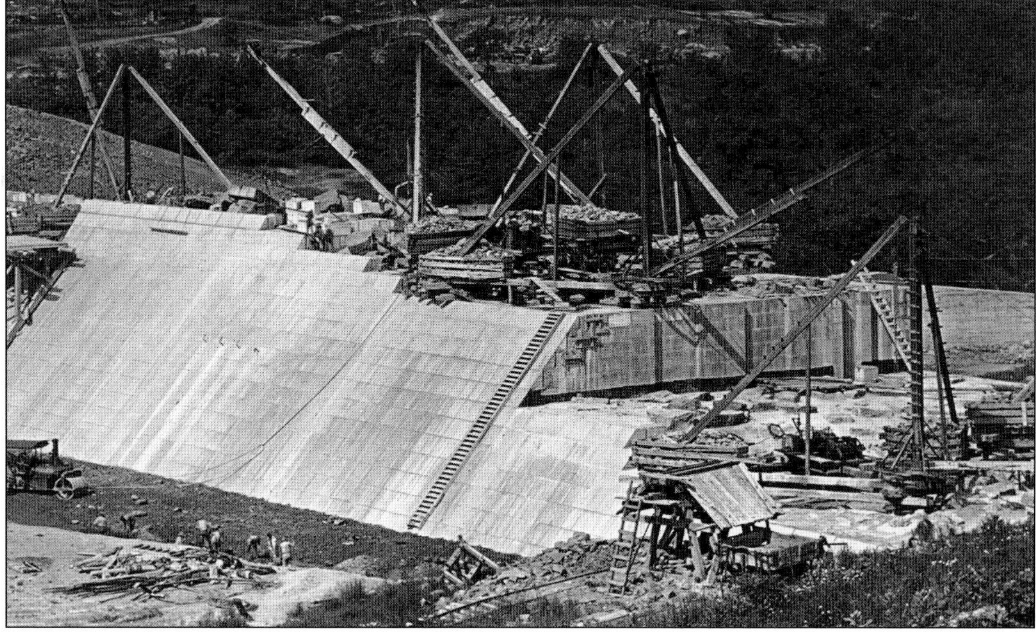

This close-up view of the north end of one of the sections shows detail of the vertical drain, behind the ladder on the right, and the lubricant (black line). An inch of thick grease is applied to the wall prior to masonry buildup to prevent bulk masonry from adhering to the wall. The upstream edge of the dam is behind the derrick tower.

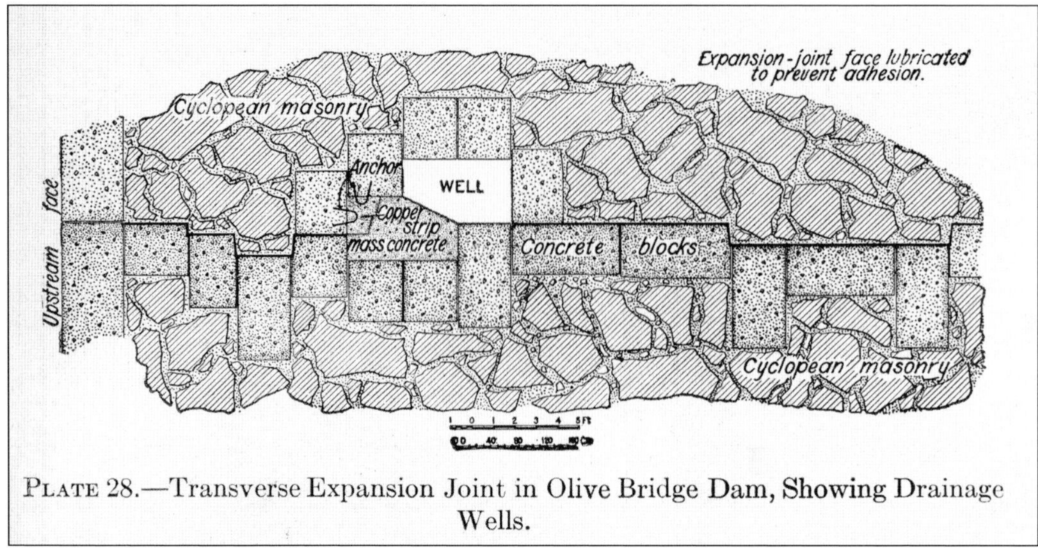

PLATE 28.—Transverse Expansion Joint in Olive Bridge Dam, Showing Drainage Wells.

This drawing shows the expansion joint designed by Waldo Smith, the first to be included in a dam. A copper plate was added during each layer of masonry as an additional water block. Reference literature stated that the completed dam had less than 300 gallons per minute of water leakage into the drain system.

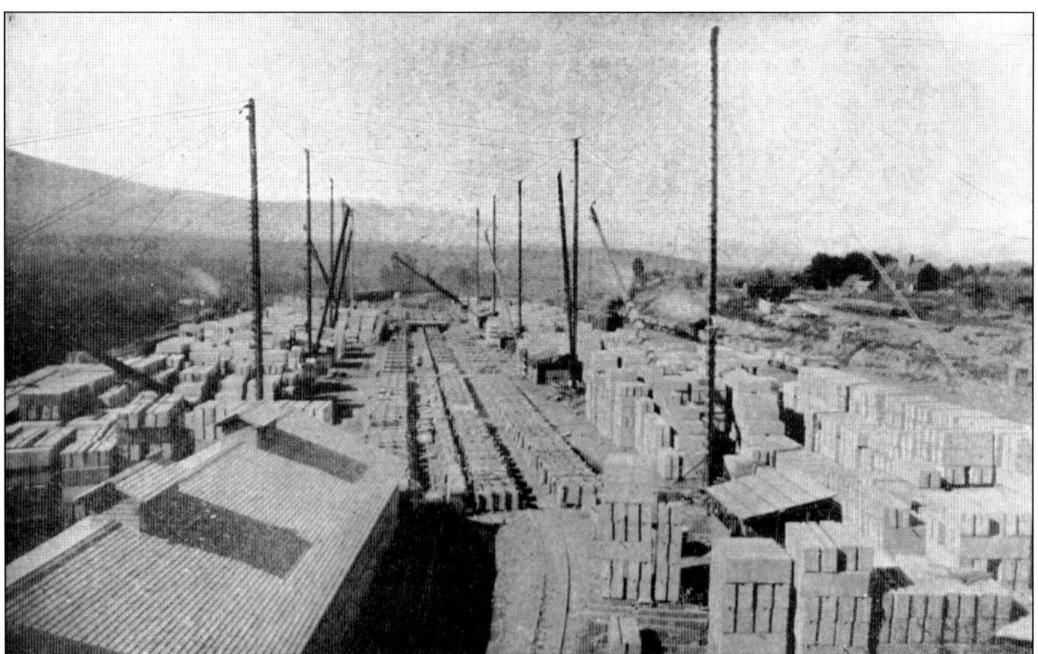

The block yard held thousands of different sizes and shapes of precast blocks. The blocks' finished sides were cast on grease-coated steel plates in forms, removed, and cured for three months before use. Blocks contained from half a cubic yard (three by three by three feet) to over three cubic yards of concrete. Fifty men worked many months casting and stacking the blocks prior to use.

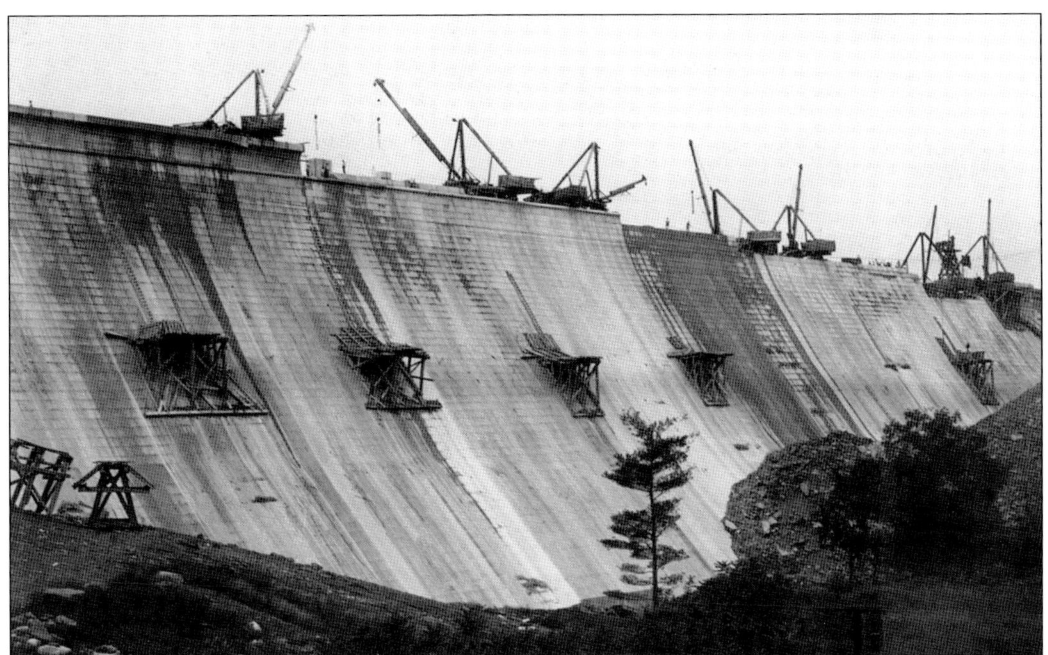

The dam is near completion. White records that in 1910, over 46,000 blocks were set as forms that contained over 425,000 cubic yards of cyclopean masonry. That is enough to cover 40 football fields with masonry six feet deep and still have some leftover. The blocks weigh between 2,000 pounds to close to 12,000 pounds (one to six tons). (Courtesy of NYPL digital library.)

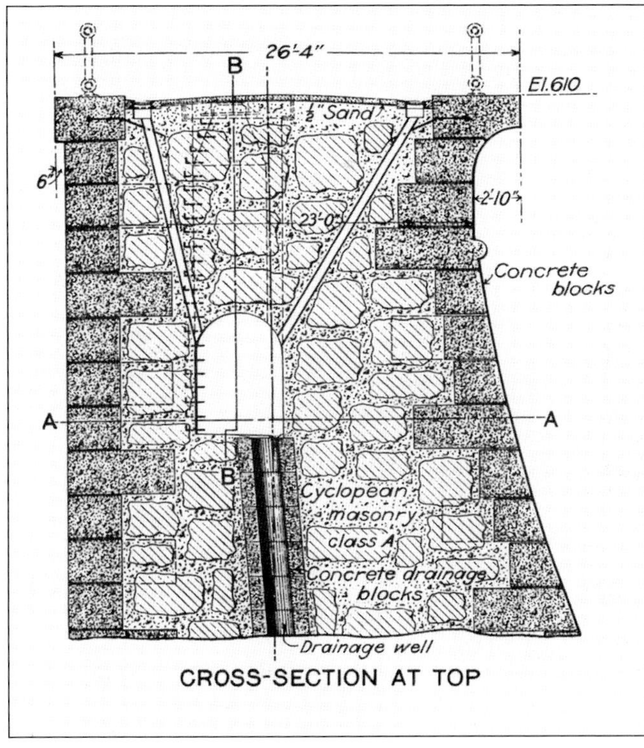

This cross-section of the top of the dam represents the final work except for railings. Until roads were needed, standard-gauge railroad tracks crossed the dam. The finished roadway is brick laid on a sand base on finished masonry. The roadway drains feed into the dam's internal drain system. The top sections of the wing dams carry identical top profiles. Inspection gallery construction is shown on page 37.

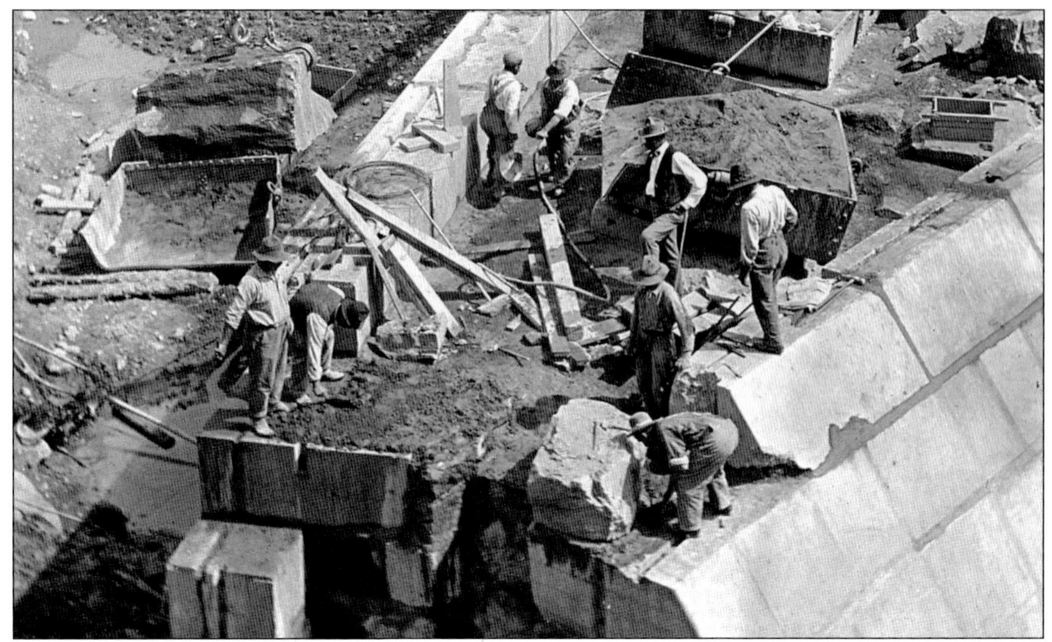

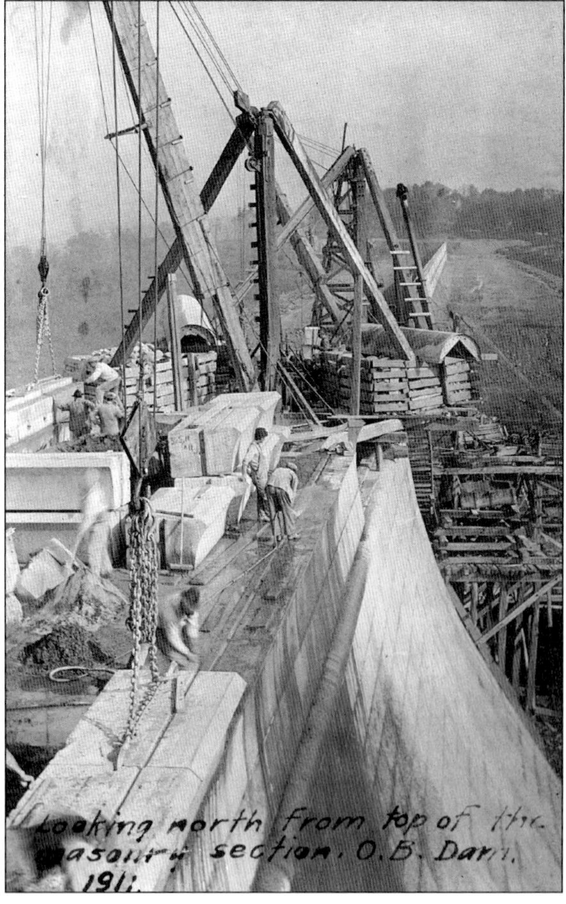

Workers are completing the south end prior to making the connection to the wing dam: a concrete core wall backfilled with standard embankments. Sub-soil on the waterside in four-inch layers was compacted to specification and then covered completely with bluestone slabs standing on edge. The opposite side was rolled to six-inch layers and finished with topsoil and grass. (Courtesy of NYPL digital library.)

Workers complete the north end of the masonry dam by adding block and more cyclopean concrete. The mating end of the north wing is visible past the derricks. Formwork and masonry join the dam and the core walls. Sub-soils were dumped, graded, raked rock-free, rolled to specification, and finished as described above.

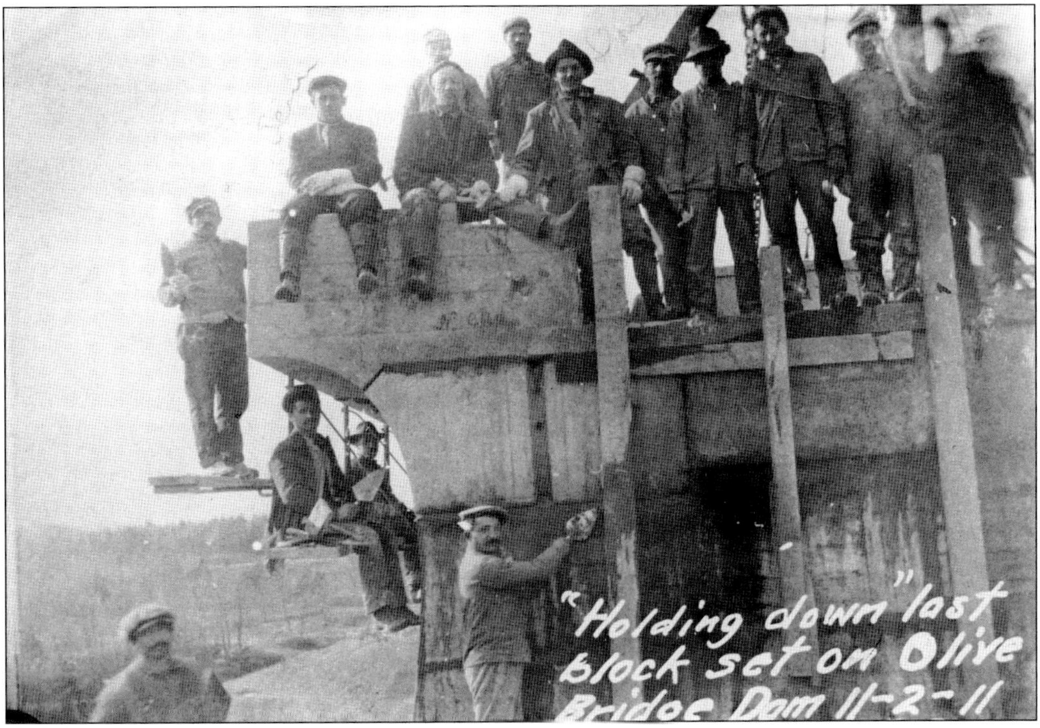

The last block, shown in this photograph, was placed and pointed on November 2, 1911. The first block was placed on the foundation on September 23, 1908. This massive 1,000-foot-long, 220-foot-tall dam was nearly completed in just over three years. The BWS year-end report listed 17,243 workers on all projects. Ashokan personnel were not separated.

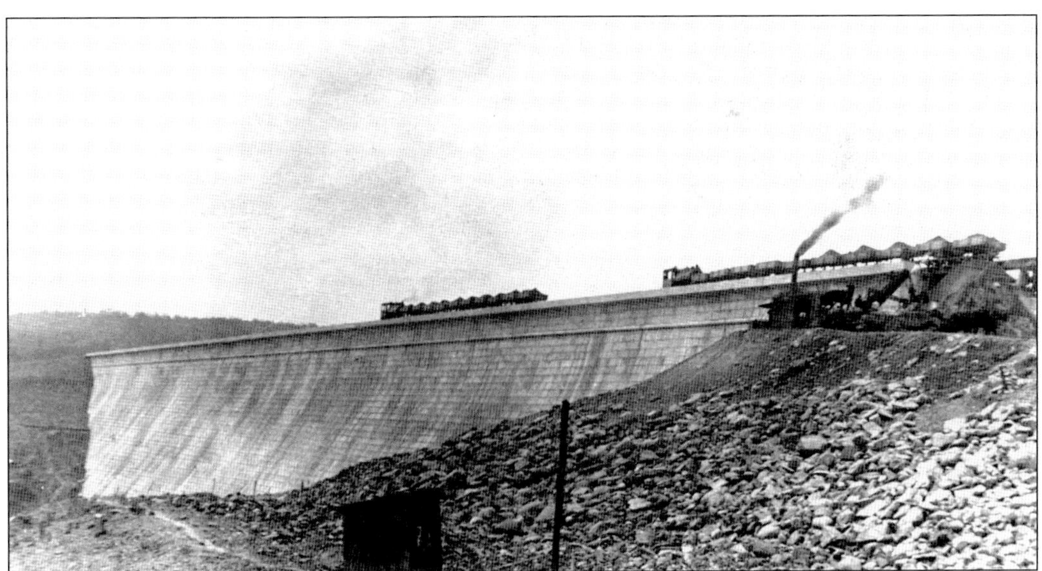

Prior to the roadway surface, railroad tracks were laid over sand on the main dam and wings to assist in placing fill on both sides of the dam. Dump cars capable of emptying to either side were used, making the task faster. Winston & Co. had 72 of these cars. Workgroups below raked the fill flat.

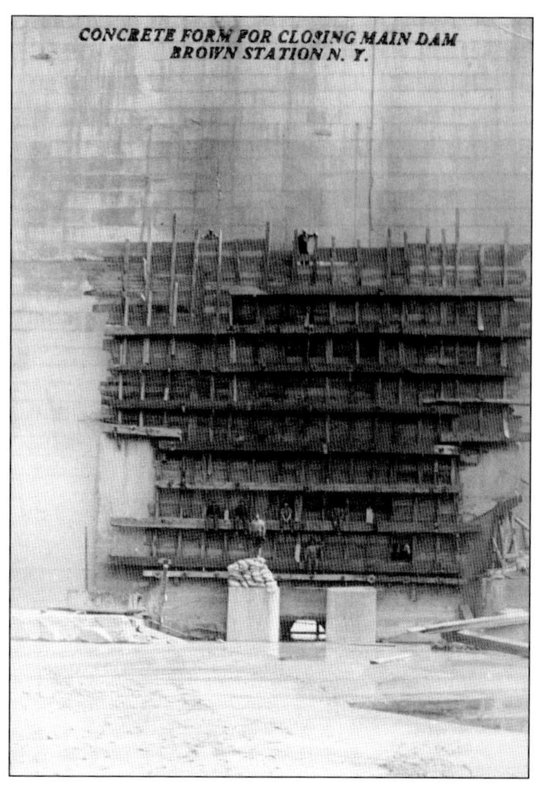

In the summer of 1913, Esopus water flow was low. The form built on the upstream dam face provided a 12-foot lap joint around the conduit opening. Twenty-four-hour work using concrete brought by mule cart, cableway, and derrick filled this form. Water flow into the 10-foot-long sluice was blocked by an upstream dam. Two square posts flanking the sluice opening held the closure boards.

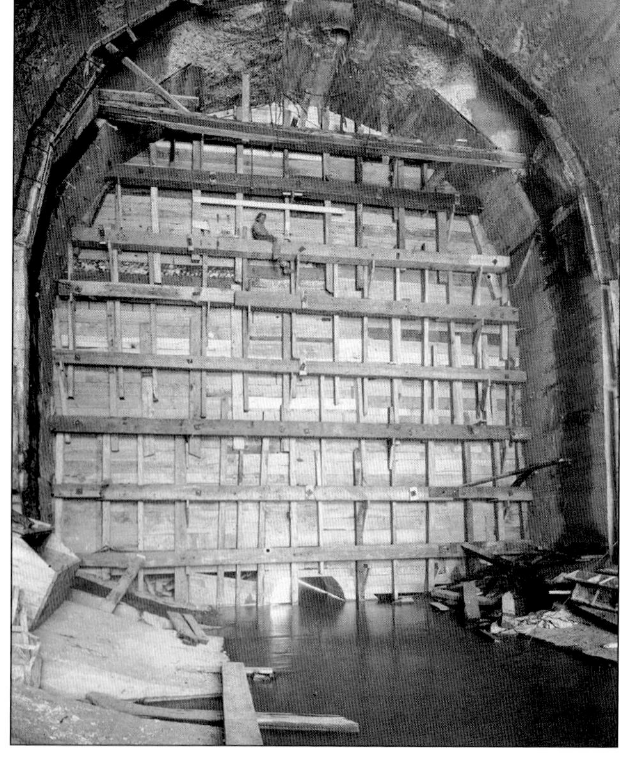

Both forms were removed after the concrete was solid. Stepped interior walls, with the narrow opening at the dam face, prevent the plug from pushing through. Workers filled the sluice's six-foot diameter, twelve-foot-long pipe with rubble and concrete. A mixing plant and rock were supplied from just outside the conduit opening. (Courtesy of NYPL digital library.)

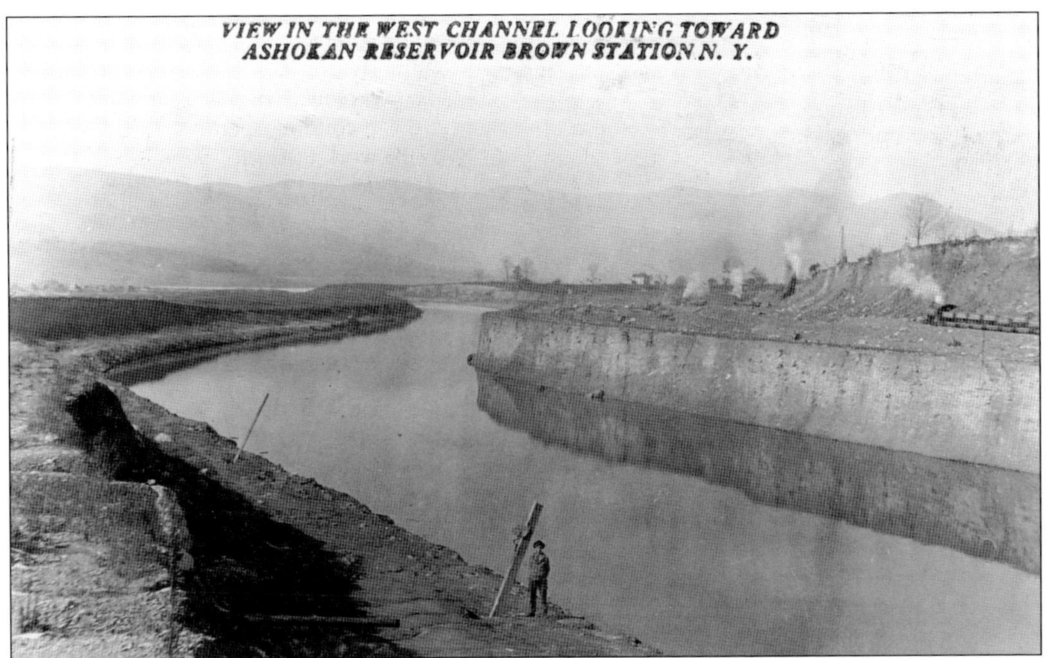

The dam was sealed on September 9, 1913, at 6:05 p.m. Smith's Annual Report states, "Sluice gate was shut. . . . Esopus Creek was flowing at the rate of 22 cubic feet per second and remained practically constant until Sept 20." Rock and cement were used to fill the sluice. Closure of the entire conduit lasted into the next year. Approximately 9,000 cubic yards of concrete were used.

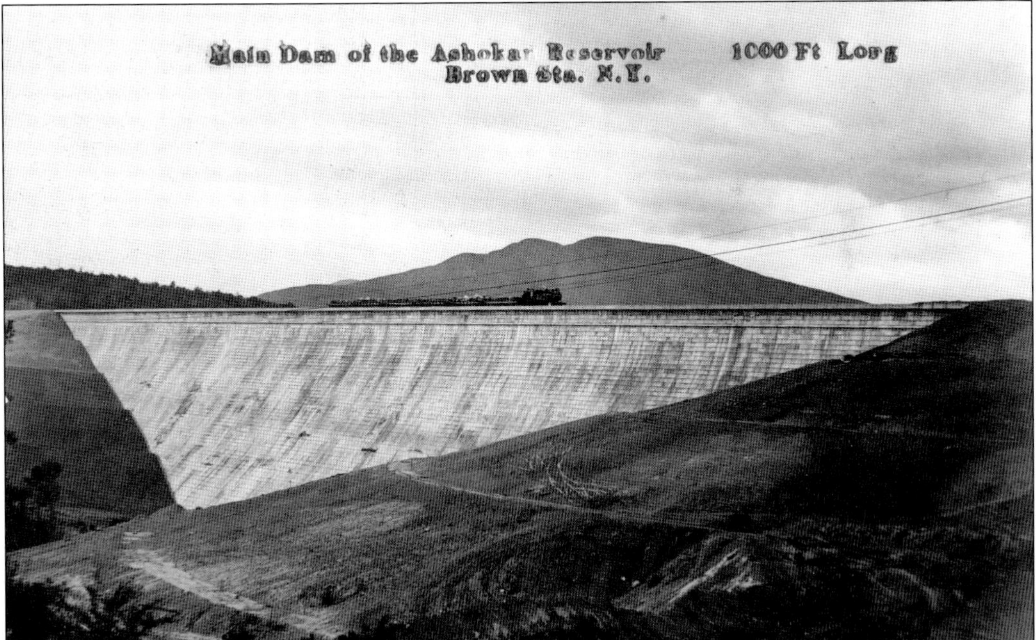

The finished dam supports a train full of Yale Quarry bluestone pieces, moving north across the dam and wings. Pieces are delivered to workers paving the waterside of the dam wings. This was a shorter path than the three-mile route over the middle of the basin. The Yale Quarry is now part of a hiking trail accessible from the southwest side of the reservoir.

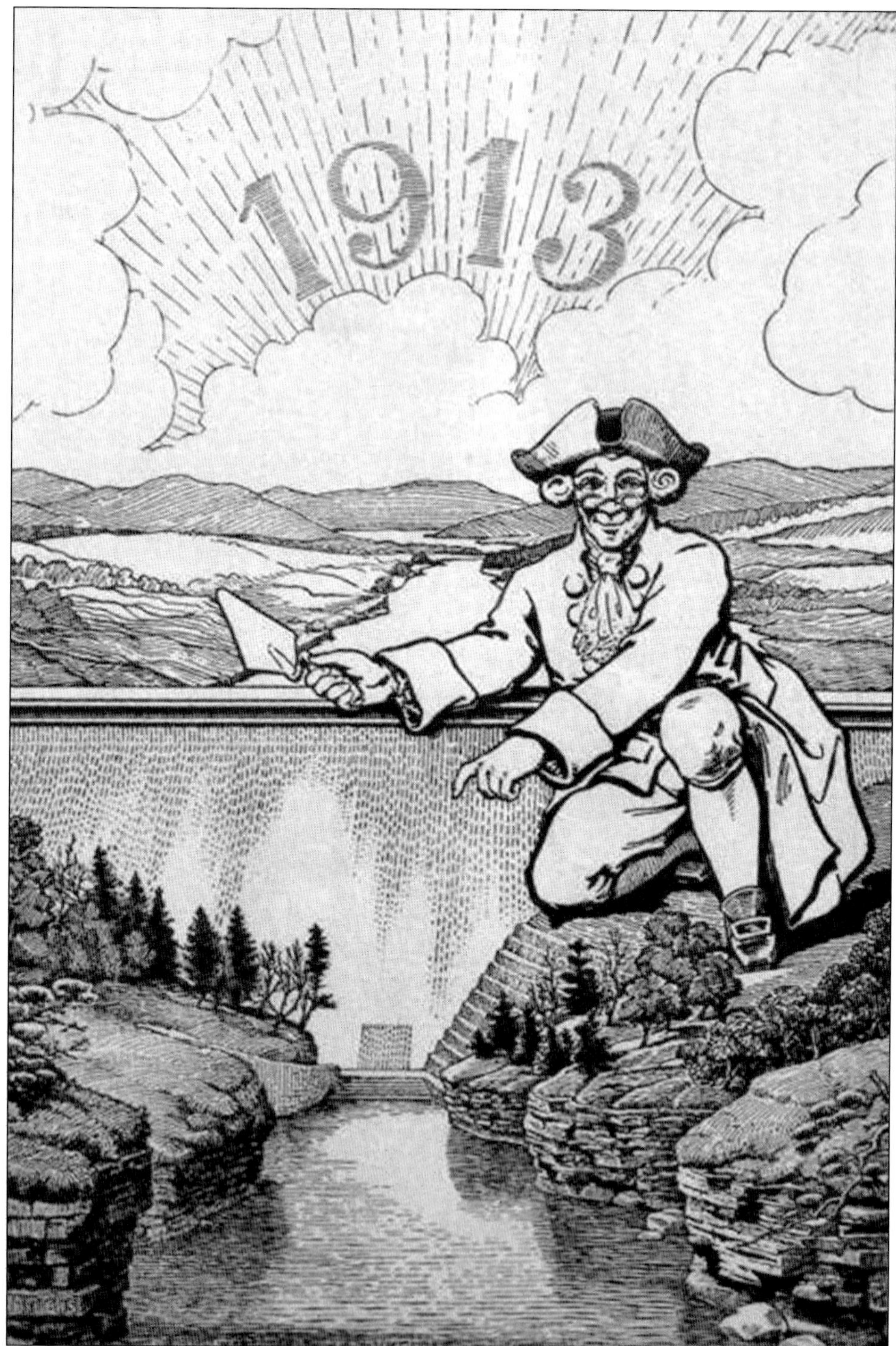

A colonial character with a trowel in hand leans on the Olive Bridge dam and points to the sealed portion of the dam that was finished in 1913, initiating water storage. This was the cover of a four-page brochure listing the menu and songs for a celebratory lunch on the dam. (Courtesy of NYPL digital library.)

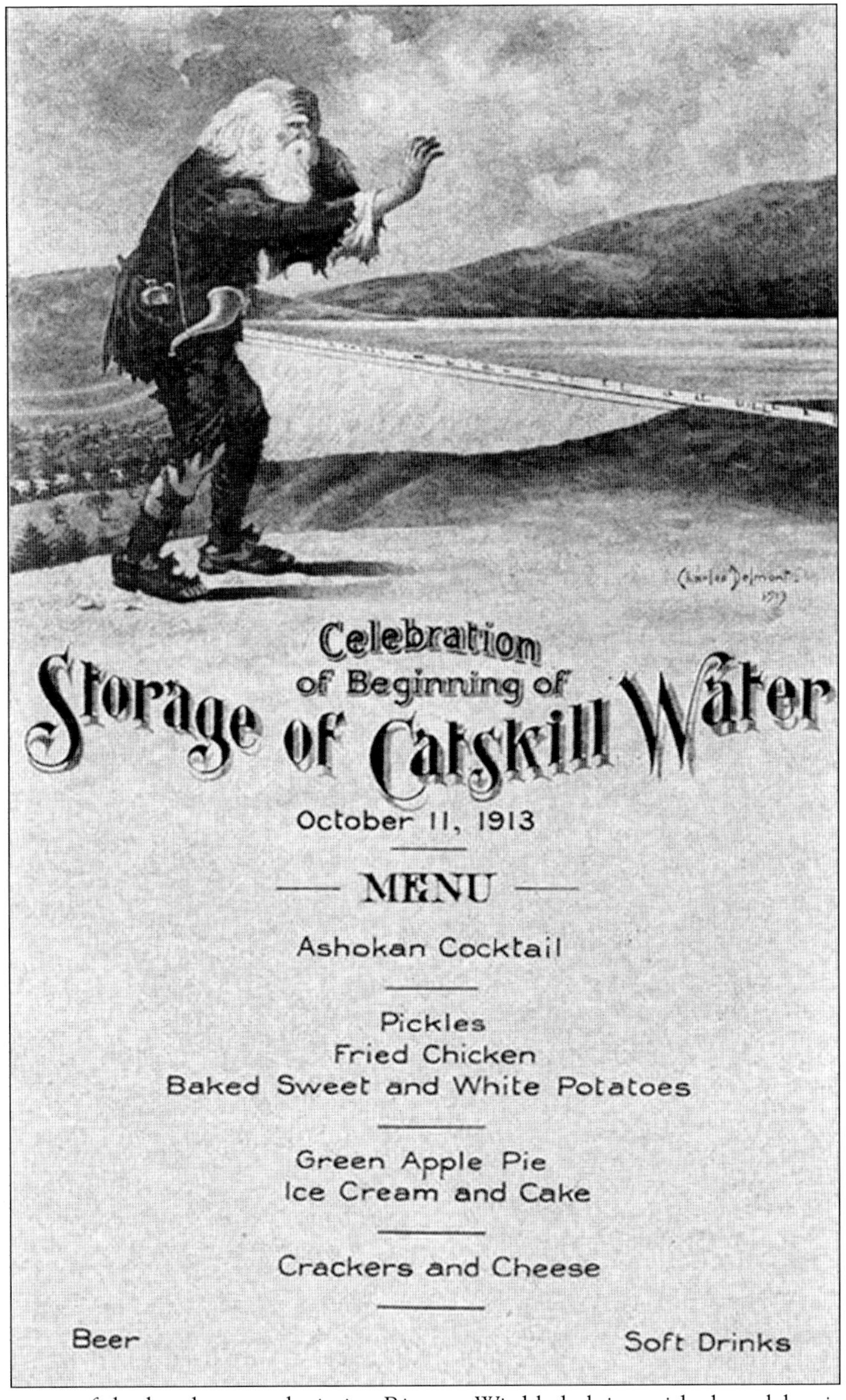

This is a copy of the lunch menu depicting Rip van Winkle helping with the celebration of the start of water storage. October 11, 1913, was the date of the celebration. New York October days can be bright and sunny, but the day of the lunch was not. The sky in this illustration appears to be grey with a threat of rain. (Courtesy of NYPL digital library.)

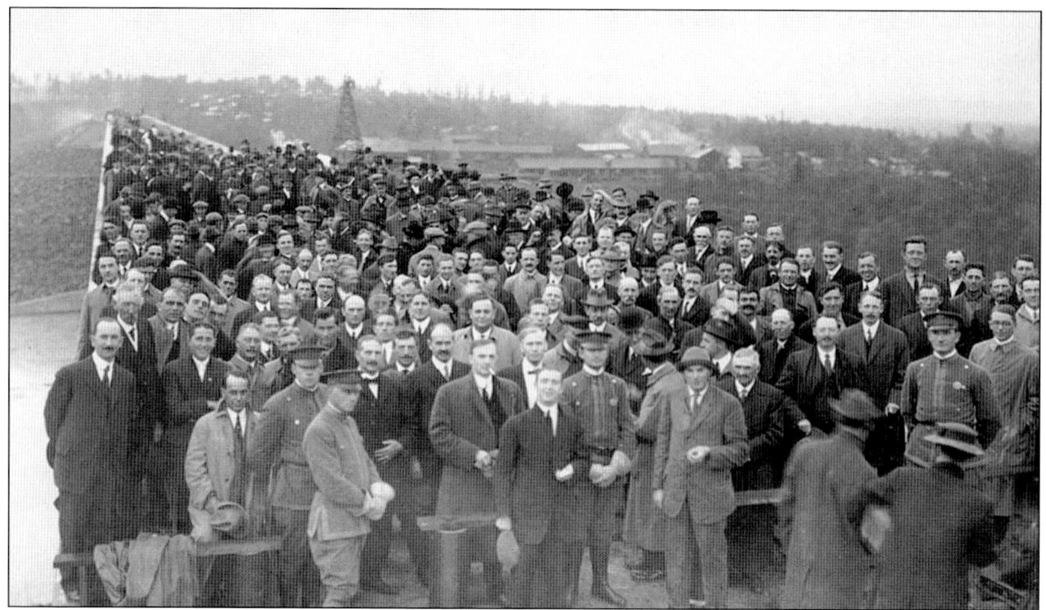

All workers were invited by the contractor for lunch on top of the dam. Guests are lining up, waiting for the signal to take their places. Those in front were probably not the men who emptied cement barrels all day. Everyone was properly dressed for the occasion, with no work for the rest of the day. (Courtesy of NYPL digital library.)

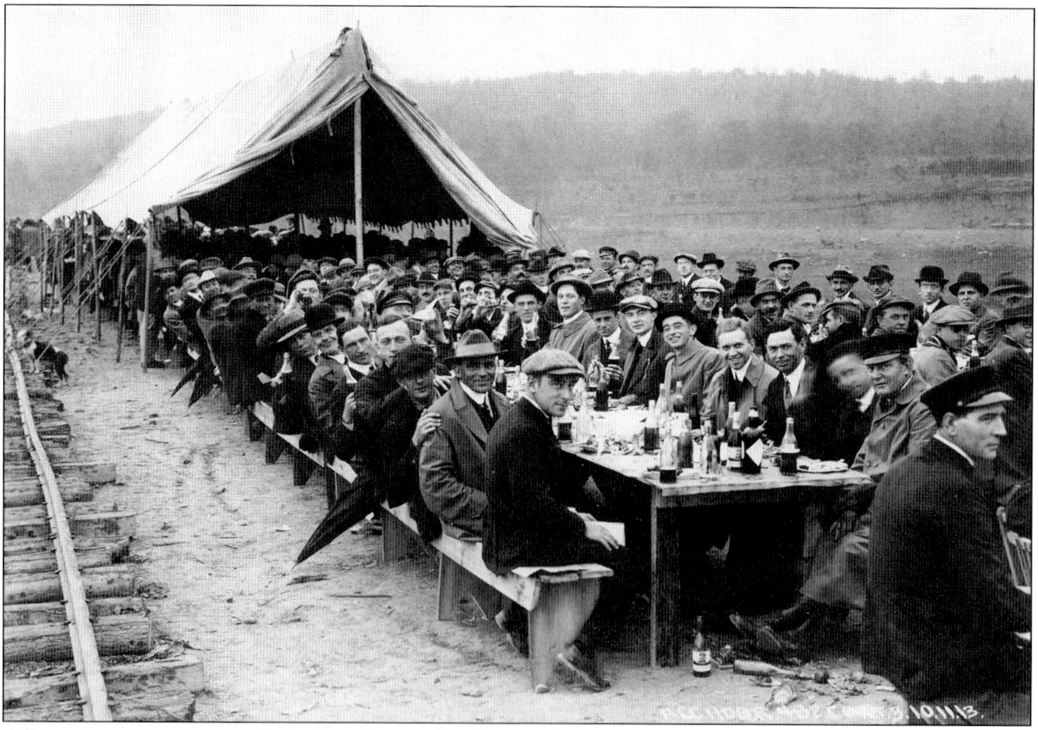

All participants are ready for lunch on the dam. The man at far right is a band member. It was not a great day for a picnic, but it was a great day to rest and take pride in accomplishments. (Courtesy of NYPL digital library.)

Three

The Dividing Weir and Water Control

By design, the reservoir was divided into two approximately equal volumes—the East and West Basins. To accommodate incoming water that may at times be cloudy with suspended clay particles and natural debris, the West Basin was the trap. It is approximately six miles long, giving water the opportunity to spread across the area, allowing particles to settle. Numbers near the top of the map on the next page show water level height. The West Basin is at 590 feet elevation, while the East Basin is at 587 feet.

Water may be drawn into the aqueduct from either basin, at different levels, or discharged from the West to East Basin through sluice gates. Four gates are at the south end of the dividing weir. Water that flows over the 15 spillways under the bridge arches is usually cleaner than water at lower depths. Intakes to the aerators and aqueduct are through the upper gate chambers, near left center on the map. There is a channel barrier wall built into the reservoir bottom to ensure that dividing weir water will not immediately flow into the intake chambers.

The intersection of the three dike roads is just above the lower gatehouse and at 610 feet elevation, the same as the top of the dam. The lower gatehouse, almost 100 feet below, controls flow to the aqueduct or aerator. The road to the left, over the West Dike, loops around the west side of Winchell Hill, the long north wing, over the Olive Bridge dam, the south wing, and to the perimeter road. The dividing weir bridge had conduits installed to allow electrical power and telephone service to the gatehouses and BWS buildings near the aerator basin.

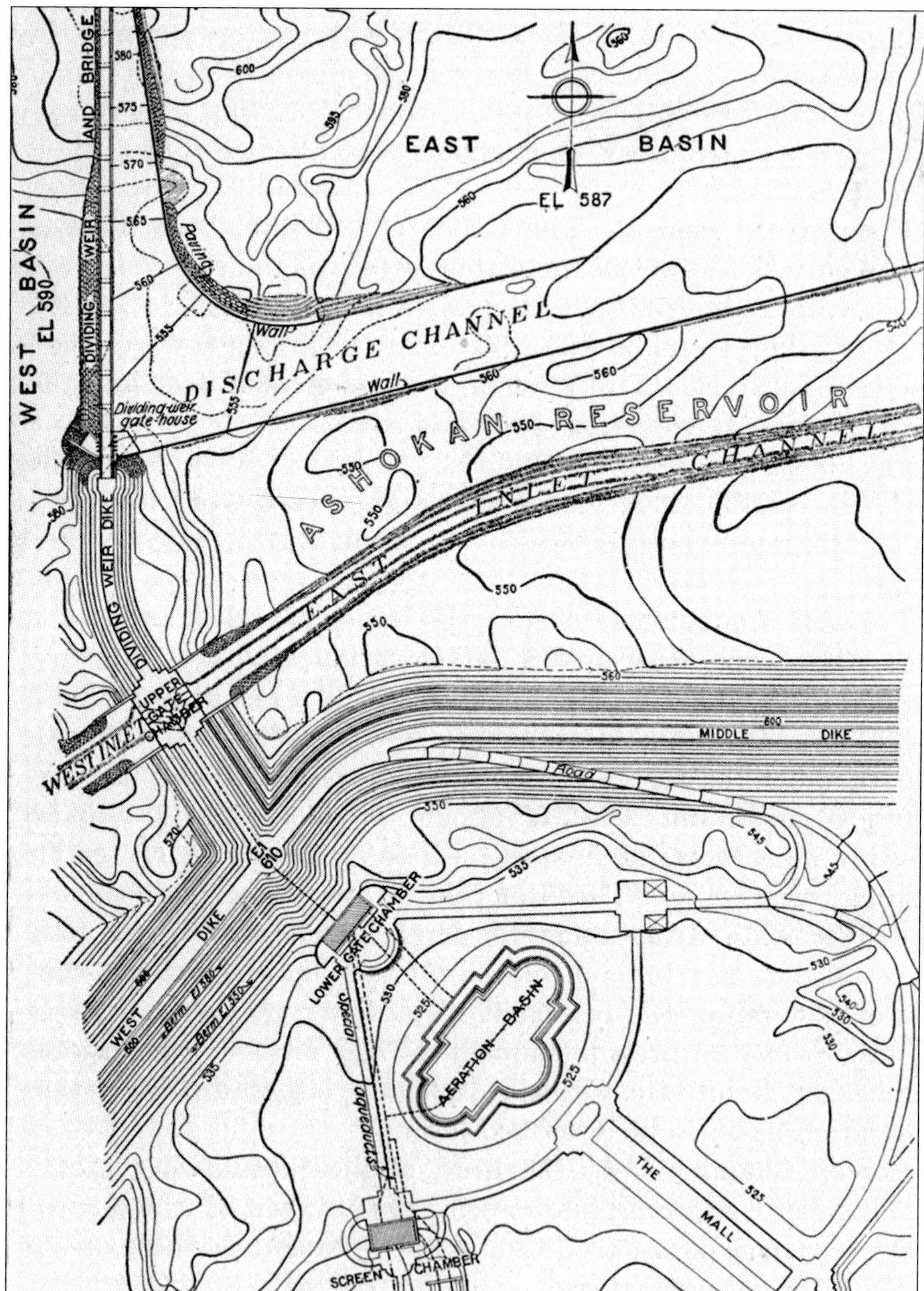

This map from White's book shows how the dividing weir, the gate chambers, and West and Middle Dikes tie together. Middle Dike was earlier named Beaverkill Dike for the Beaverkill streambed under the dike. Also shown are the underwater channel guides.

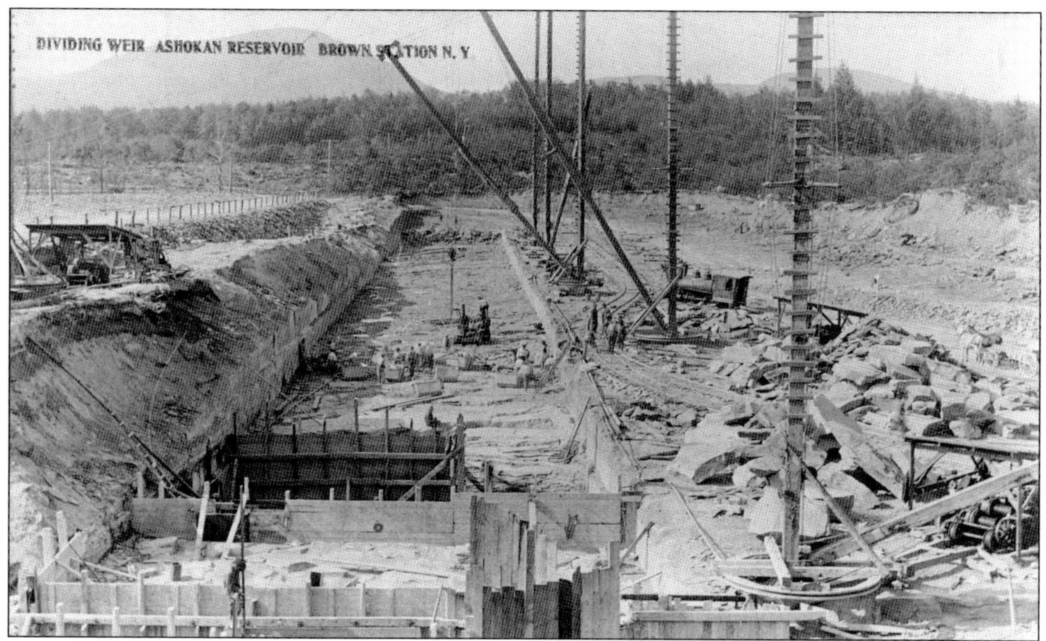

This view shows the 1,000-foot base for the dividing weir with the sluice gates support in the foreground. Soil was removed to bedrock. All loose, fractured material was removed to sound rock to allow a good connection with concrete. All concrete used in the weir was from the main mixing plant and delivered by cableway.

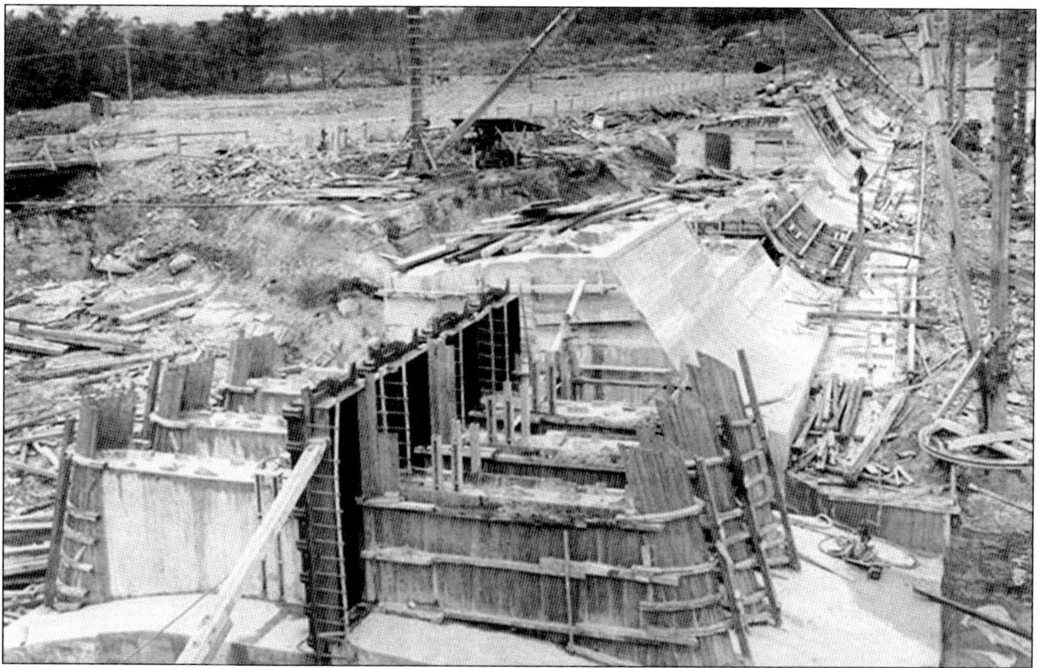

This view shows four sluice gates in place as well as their inlet and outlet separation castings. Sluice gates allow a rapid movement of water from the West into the East Basin. There are no other means of water transfer other than overtopping the 15 spillways. Work progressed from the south to the north end. (Courtesy of NYPL digital library.)

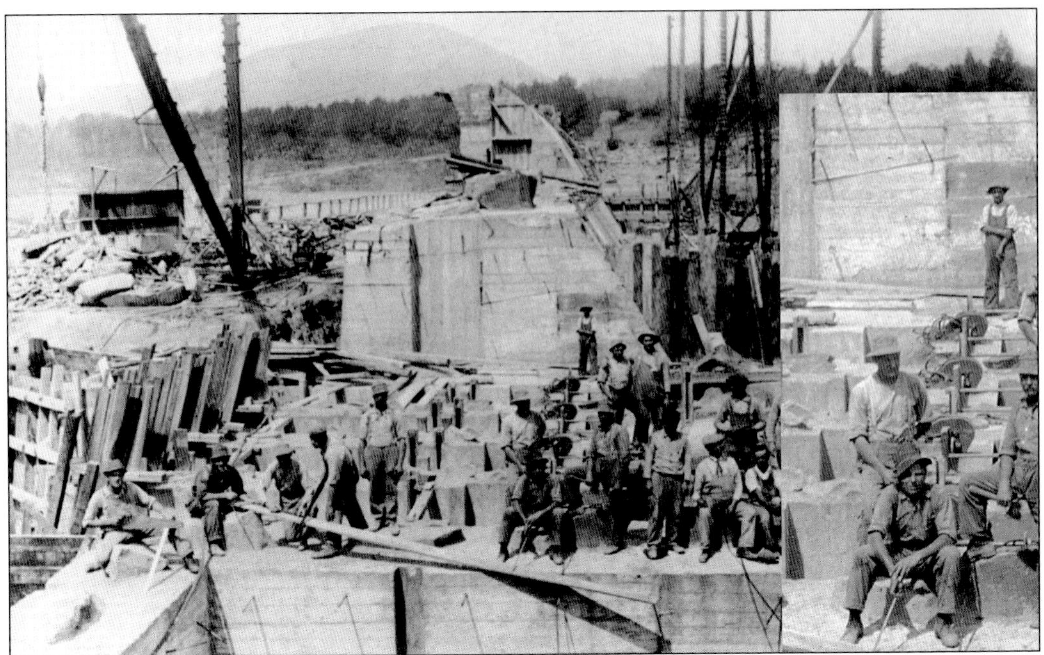

The crew is on a break with one of the four sluice gate wells in front and the north wall of the gate chamber behind. The upper curve of one of the cast spillways is behind the workers. Photographic film of the early 1900s required a long exposure. Subjects had to remain still for at least 30 seconds to avoid blurring.

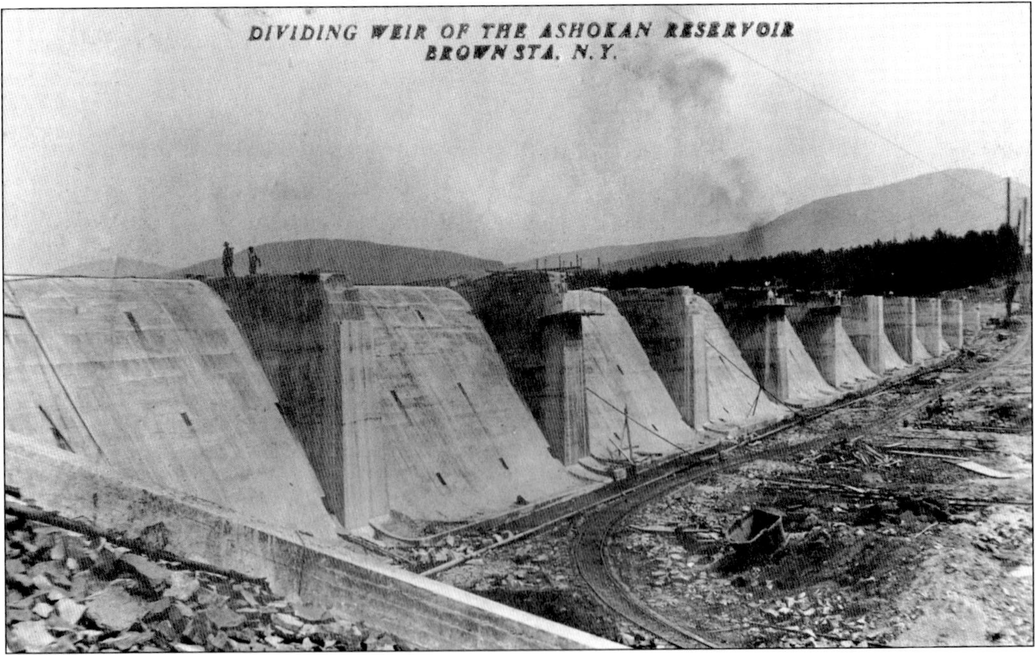

This view of the less-than-halfway completed weir shows the spillway and bridge support casting essentially done, with the spillway arches and roadway structure yet to be added. At left is the southern wall of the sluice and spillway discharge channel. The narrow-gauge tracks were used by trains delivering materials and workers.

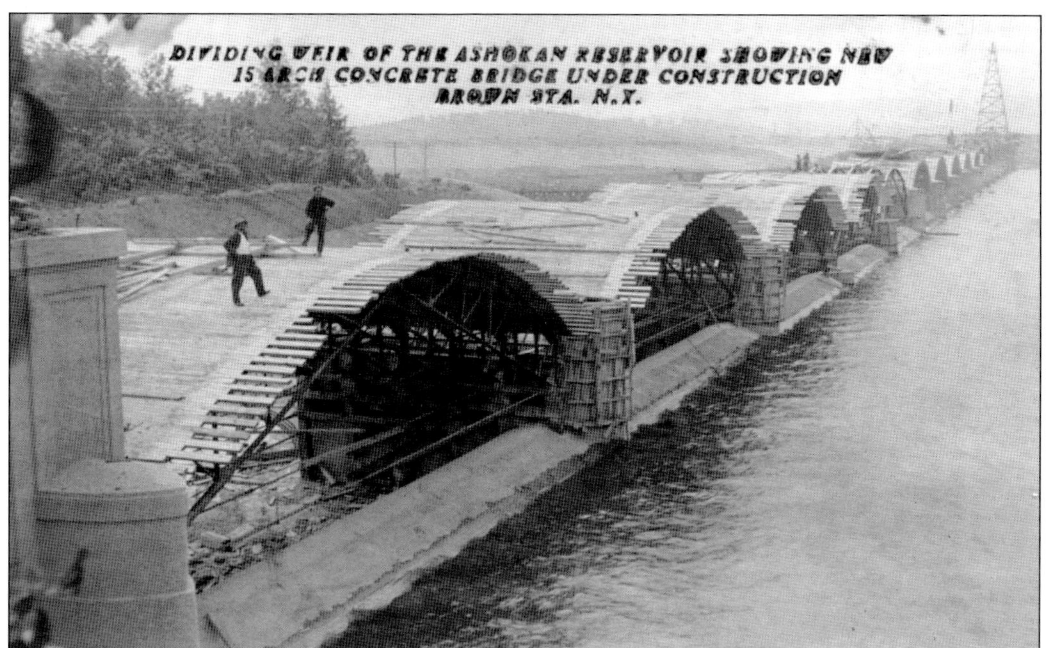

This view from the north end of the near-full West Basin shows wood forms over cribbing that supports the roadway masonry. Roadway supports and arch casting still needed to be done. Concrete was transported to the forms using the cableway, as it was for the dam. The tower in the distance is the cable return tower.

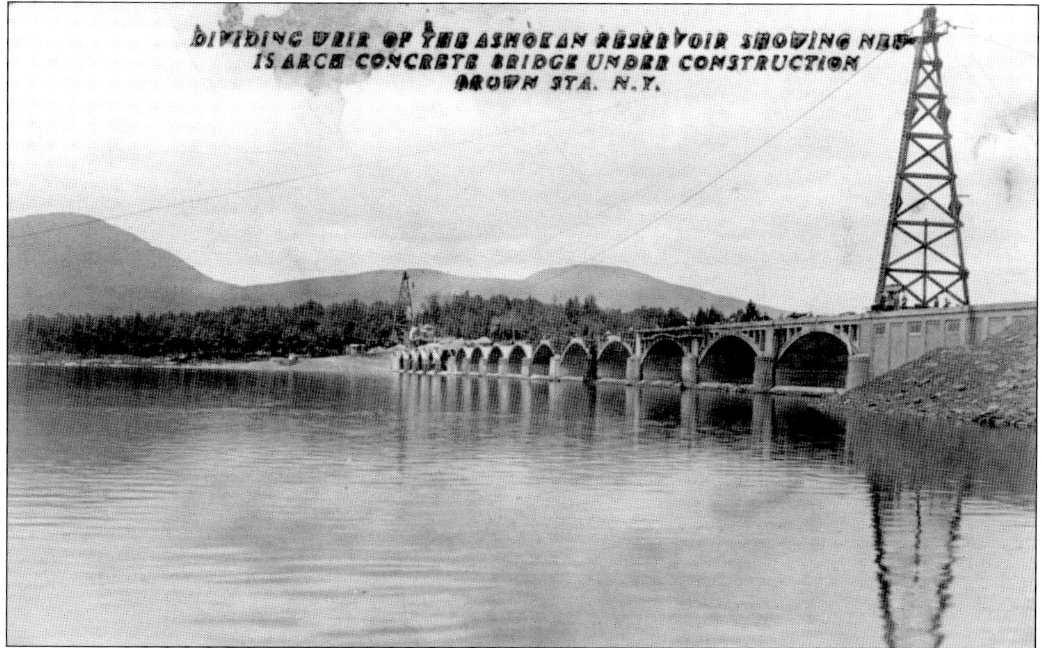

This view, probably from 1911 or 1912, looking north, shows the West Basin nearly full of water. Roadway work continues. The roadbed over the rightmost arch is complete. The cableway would remain until the roadway was finished. Plans are underway to rebuild this now 110-year-old structure into a bridge that has a wider roadway, including bike paths, and a larger load capacity.

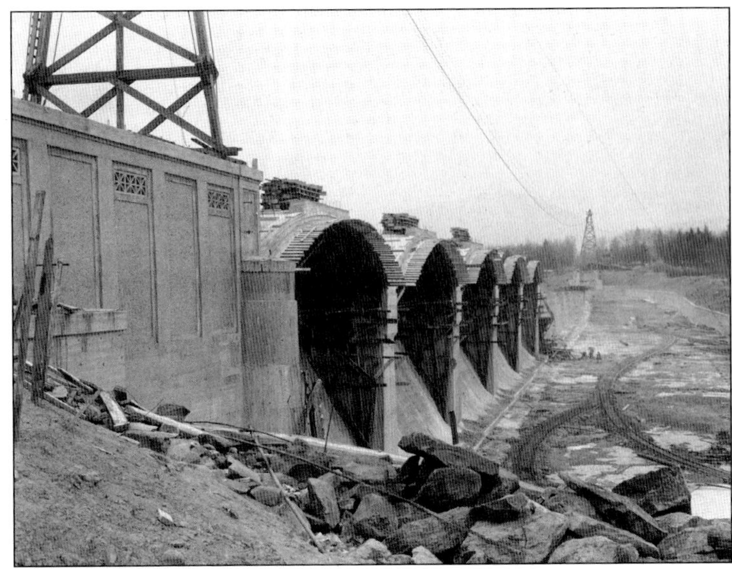

Beginning masonry casting is underway on the first four arches. Forms have been delivered for additional masonry to support the roadway. It seems that these forms were delivered after the work crews finished for the day. A similar practice was done on the main dam. Materials needed for daylight work were delivered during the night, so the cableway was free to deliver concrete. (Courtesy of NYPL digital library.)

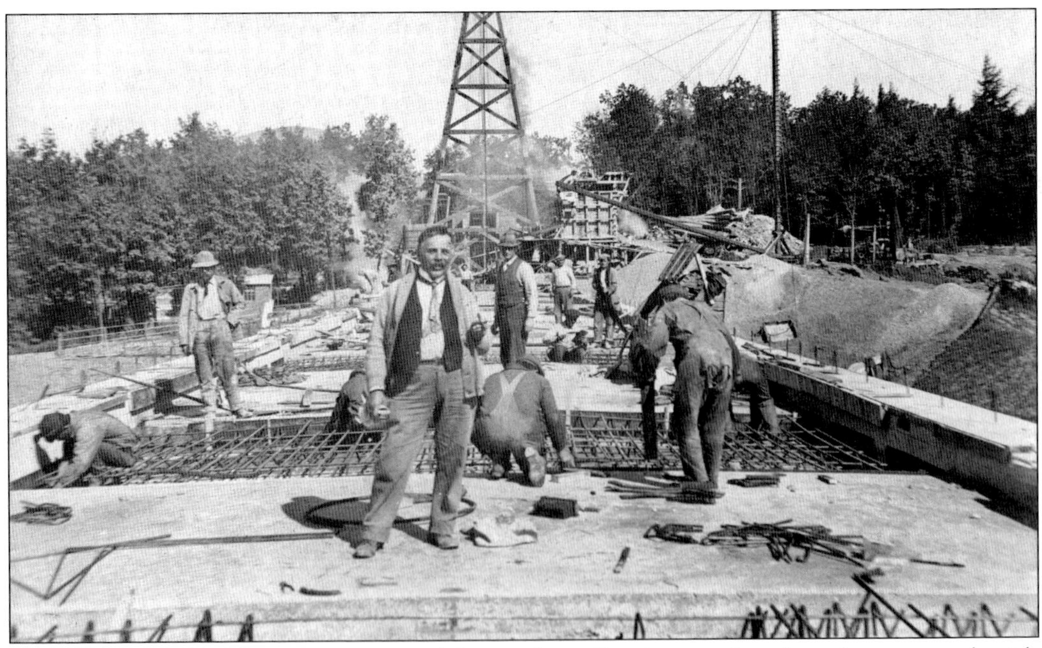

This crew is casting alternate sections of the roadway. Previous sections have been poured, with reinforcing rods extending into the new sections. Rod extensions are wire tied to the rod in the work area. Concrete was delivered into each section, making a complete roadway. The men in vests were probably BWS inspectors or engineers.

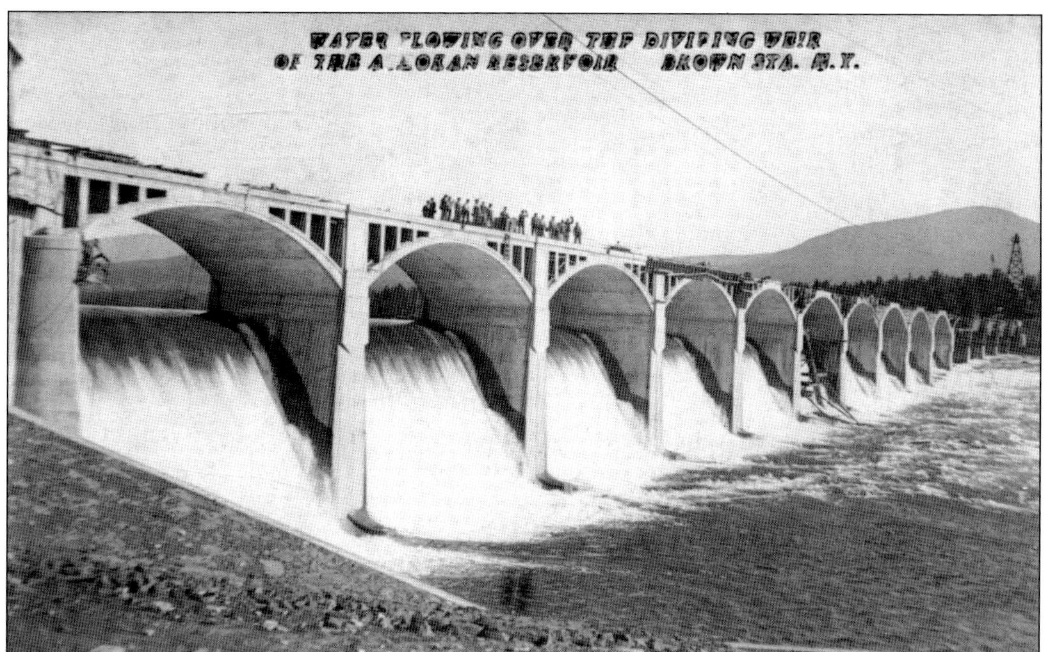

Roadway work was underway when heavy rains caused the West Basin to overflow. Formwork in one of the arches appears to have been damaged by the heavy water. Note how the spillway design directs the water smoothly into the East Basin discharge channel. The workers on the bridge may have been wondering what the foreman had in mind for the day.

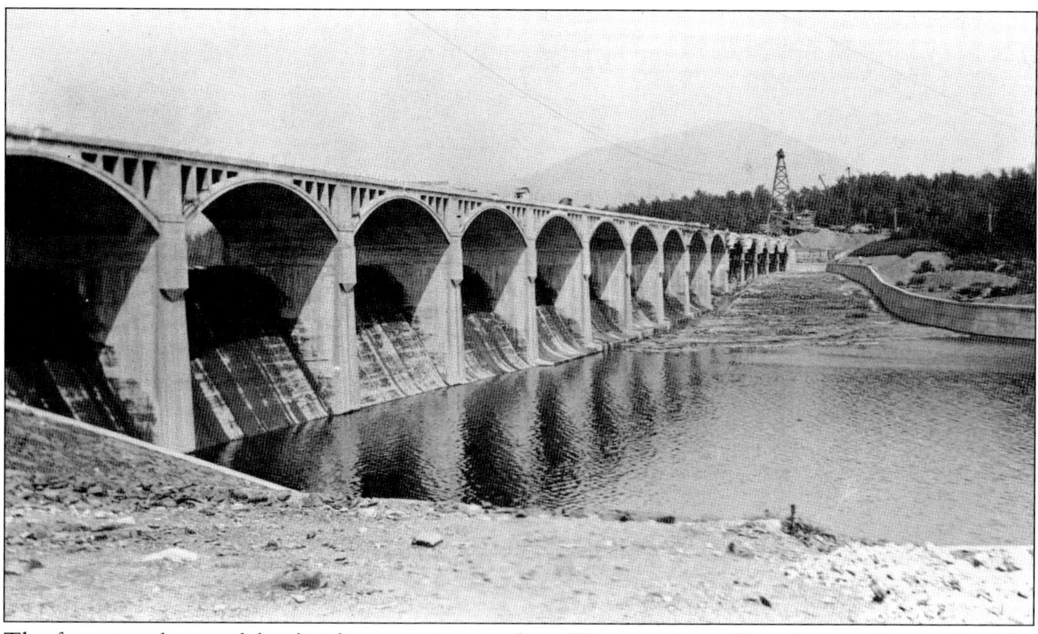

The functional part of the dividing weir is complete. Water at the spillway base indicates the dike work around the East Basin had been brought to the level that water can be contained. Water in the West Basin had not yet reached a level where it could spill over into the East Basin. Today, a fence is on top of the retaining wall on the right, making it difficult to fish.

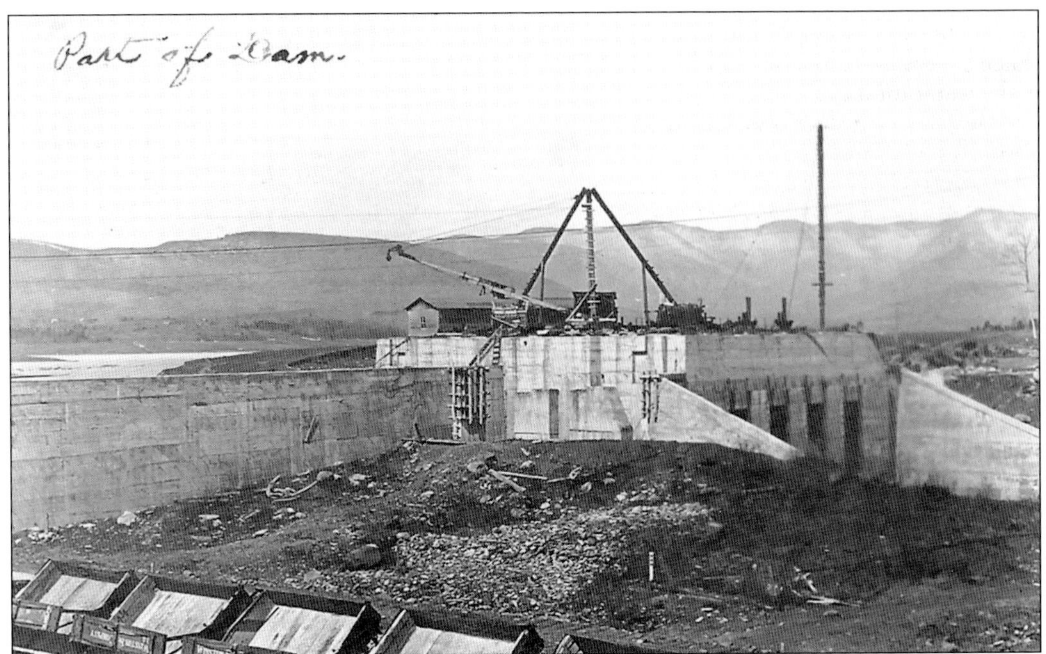

This overview image shows the intake chamber and the West Basin beyond. This important housing would eventually contain 16 intake valves. They are arranged in four sets of valves on four different levels that open into both basins. This is the beginning of the journey from water storage to water shipment, all gravity fed, to consumers in New York City.

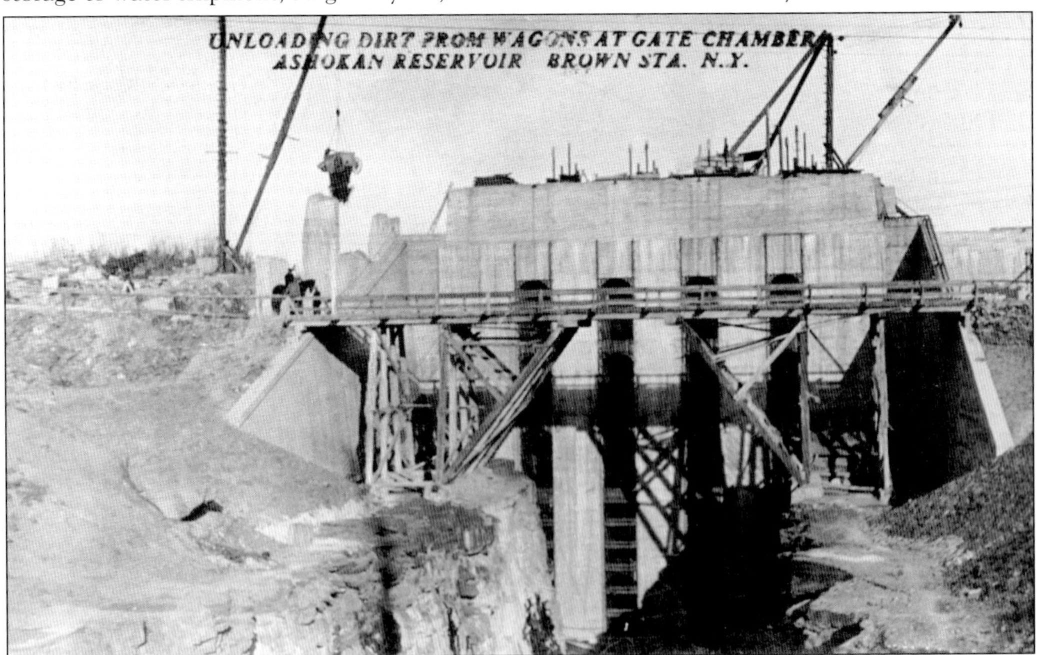

In a very brave move, this airborne wagon with a load of hardpan soil was dropped against the dike wall. Both surfaces are banked with packed earth and then waterside paved with bluestone pieces to 10 feet above the high water line. This view shows the West Basin side of the openings into each of the sluice gates. Bedrock was cut deeply to make a level bottom to the lowest inlet level.

A few hundred feet to the right of the sluice gate chamber over the short dike is the water intake structure. Water flow into the aqueduct is managed through valve chambers in each basin. Both chamber water inlets are identical. The West Basin outer face is shown here. A mile-long channel from Esopus Creek meets the face. Behind each opening is a pipe and valve to a common manifold and the aerator.

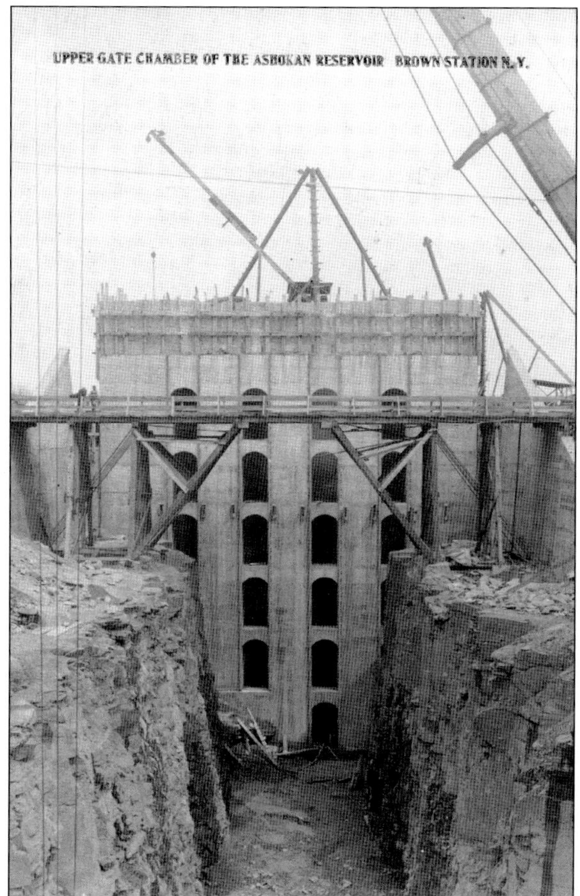

Gravity pulls water into the aqueduct through banks of four intakes. Control of the four sets is located above the building. A duplicate is on the opposite side of the weir. Each set feeds into upper and lower aqueducts. A 100-foot head feeds the lower gatehouse 600 feet away. The intakes and water delivery pipes are cast in concrete that fills the voids of the structure.

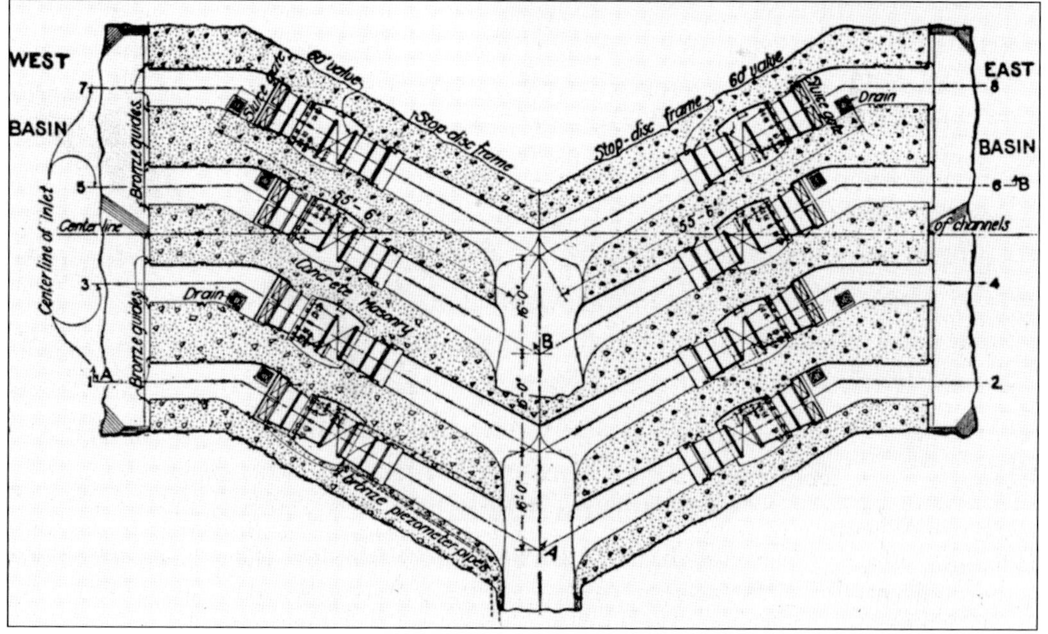

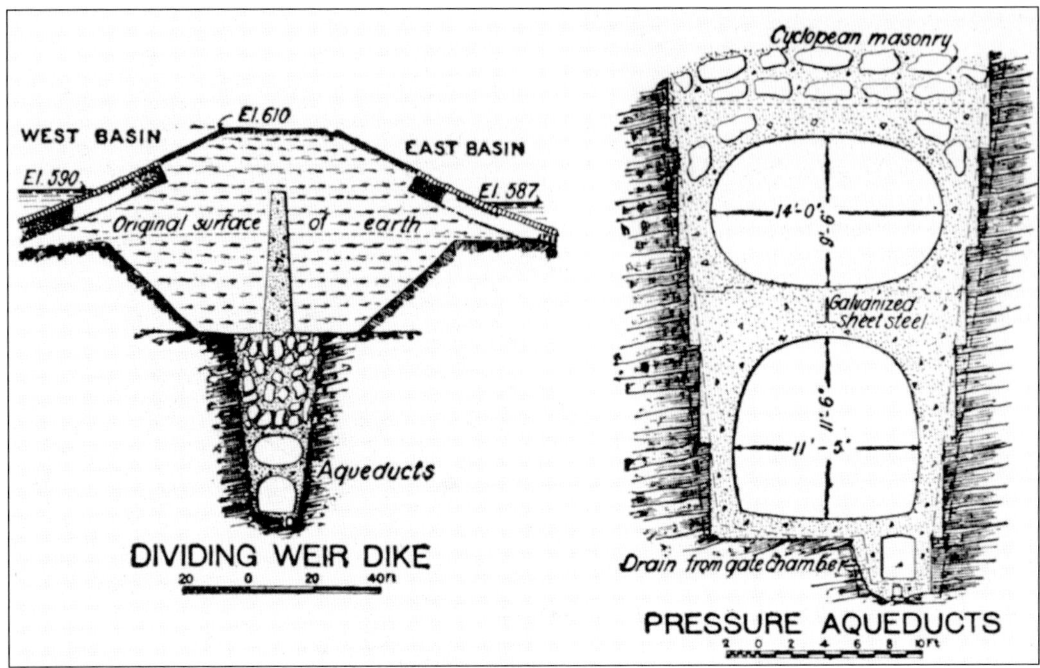

The drawing at left shows a cross-section of the dividing weir and the buried aqueducts from the dual gatehouses to the lower gatehouse. The top two intake rows feed the upper tunnel, while the bottom rows feed the lower tunnel. The horseshoe tunnel is at 492 feet elevation and is 11.5 feet high and wide. The oval tunnel is at 506 feet and is 9.5 feet high and 14 feet wide.

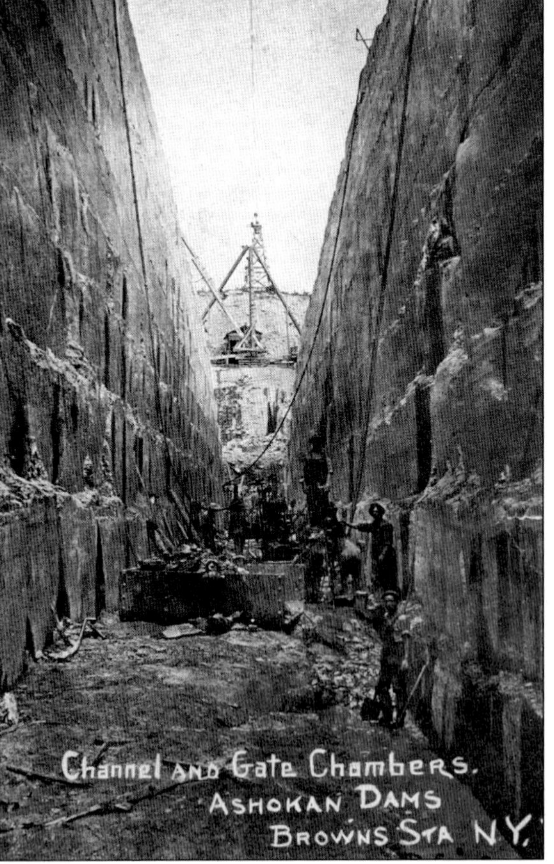

Completed trench excavation provided space for the aqueduct cross-section, shown here on the right. Water flows through either or both aqueducts from the gatehouses to the lower gate chamber and onto the Kensico Reservoir. Cyclopean masonry fills the trench that ends at the bottom of the Middle Dike, with pressure pipes to the lower gatehouse.

Most lower gate chamber views are too busy to describe. In this drawing, water moves from the dual aqueduct chamber at upper right into gate valves similar to those shown below. Water is directed into the aqueduct (upper left), or to the aerators (lower right bank of pipes), or both. Water pressure into the lower gate chamber is approximately 6,000 pounds per square foot.

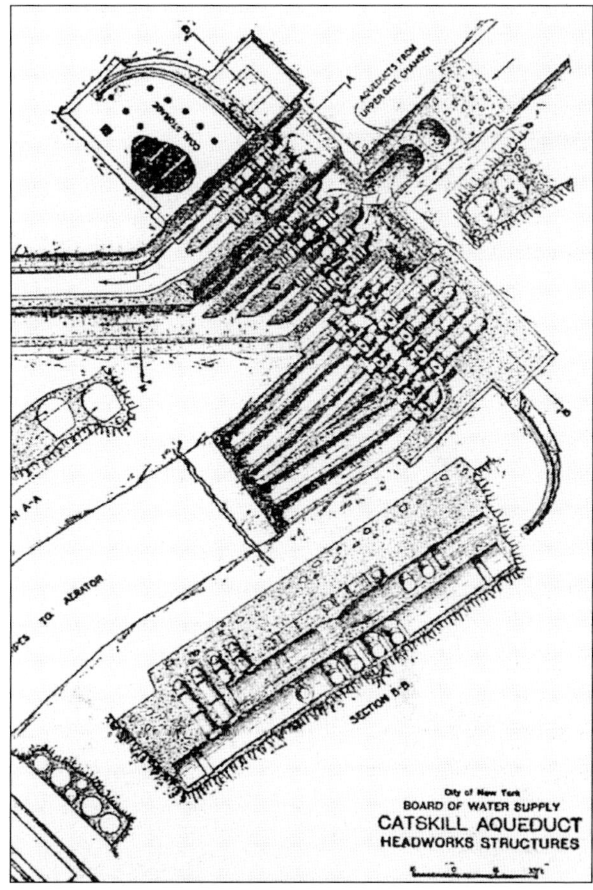

Large gate valves are the main controls of water flow to the aerator feed (right) or to the aqueduct straight ahead. Two workers are doing finishing work with the extra-large valves, one of which has at least a 72-inch opening. Valves and large connector pipes were in place before the structure was built, covering the work.

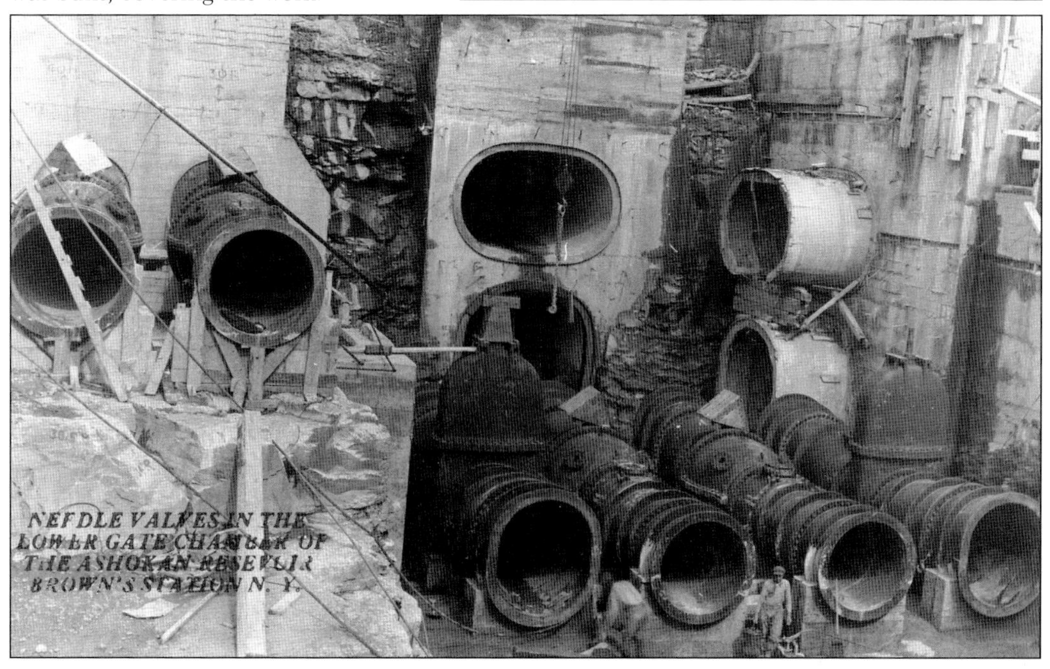

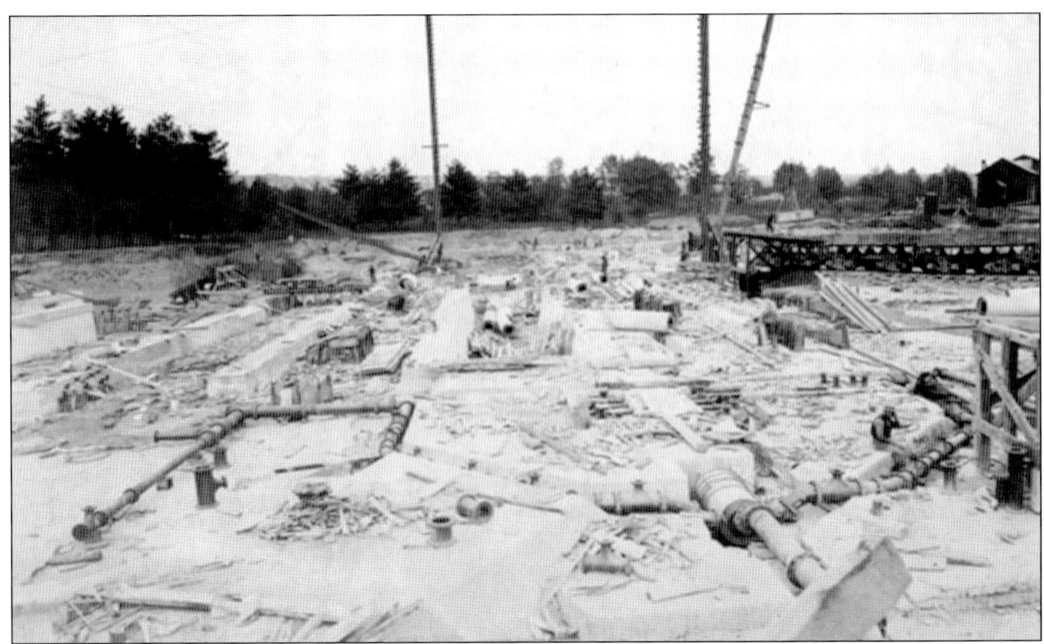

This early view of the Aerator Basin shows many small connector pieces placed near where they would be installed. This aerator work was under a different contract using tradesmen with different skills. Approximately 1,600 water jets were placed in the large oval basin that collects water after being sprayed into the air.

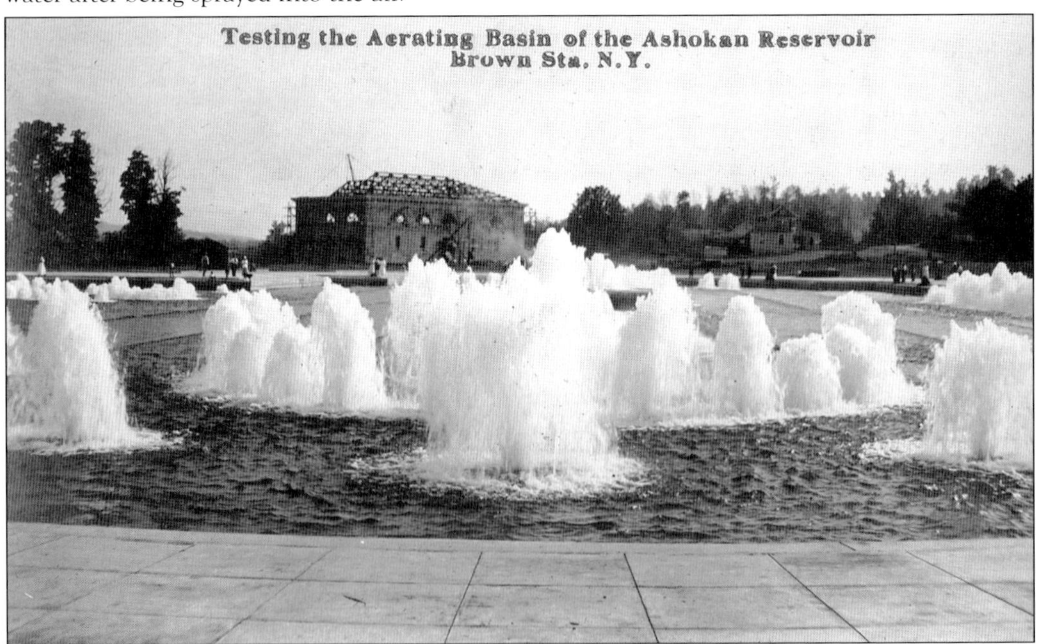

Aerator jets are tested for balanced flow. It was thought that spraying water into the air prior to aqueduct travel would improve its taste and destroy bacteria. The background building is the lower gatehouse; the piping was installed first, and then the structure was built over the work. A chlorination system later installed at a downstream reservoir eliminated the need for the aerator system. The aerators were replaced in the 1980s with a single-nozzle fountain.

Four

Perimeter Dikes and the U&D Railroad

As the dam and dividing weir progressed, work on the dikes began. The Ulster & Delaware Railroad that connected the valley communities had to be relocated. To accommodate existing operations, dikes had to permit railroad access. New York state law prohibited disturbing a railroad right-of-way until damage claims were paid in full. In 1910, the U&D claimed the total relocation cost to be nearly $4.6 million for real property lost, new right-of-way purchased, damages, and right-of-way construction. An agreement between New York City and the U&D came on June 15, 1911, for $2.8 million, with relocation work done by the dam contractor, paid by BWS. The U&D made significant money moving cement, coal, supplies, and equipment for the reservoir. Telephone lines were also moved to the north side of the reservoir.

Dike construction started in early 1909, with West, Middle, and West Hurley Dikes providing railroad portals. Earth removal was necessary, as specifications required a clean, solid bedrock base for core walls. Dike embankments started from bare hardpan earth using rock-free compacted earth banked to core walls. Cleared distance away from the core wall was three feet for every foot of dike height. Core walls were built to 590 feet above sea level water height. Backfill soil was delivered by narrow-gauge rail, dumped, and then raked by hand to a four-inch depth waterside, six-inch exterior side. Stones were moved to the outer edges. The soil was compacted after each completed layer. The finished material was brought to the height of the main dam, at 610 feet elevation, with a 34-foot-wide level surface.

When completed, the West, Middle, and West Hurley Dikes were made for two-lane auto traffic. The Glenford and Woodstock Dikes had to accommodate the U&D mainline. All interior wall embankments were paved with bluestone pieces delivered by rail and set at the embankment bottom 90 degrees to the fill plane and 10 feet above the high water line. Outer dike walls were seeded with grass. The U&D began operation on the new right-of-way on June 8, 1913.

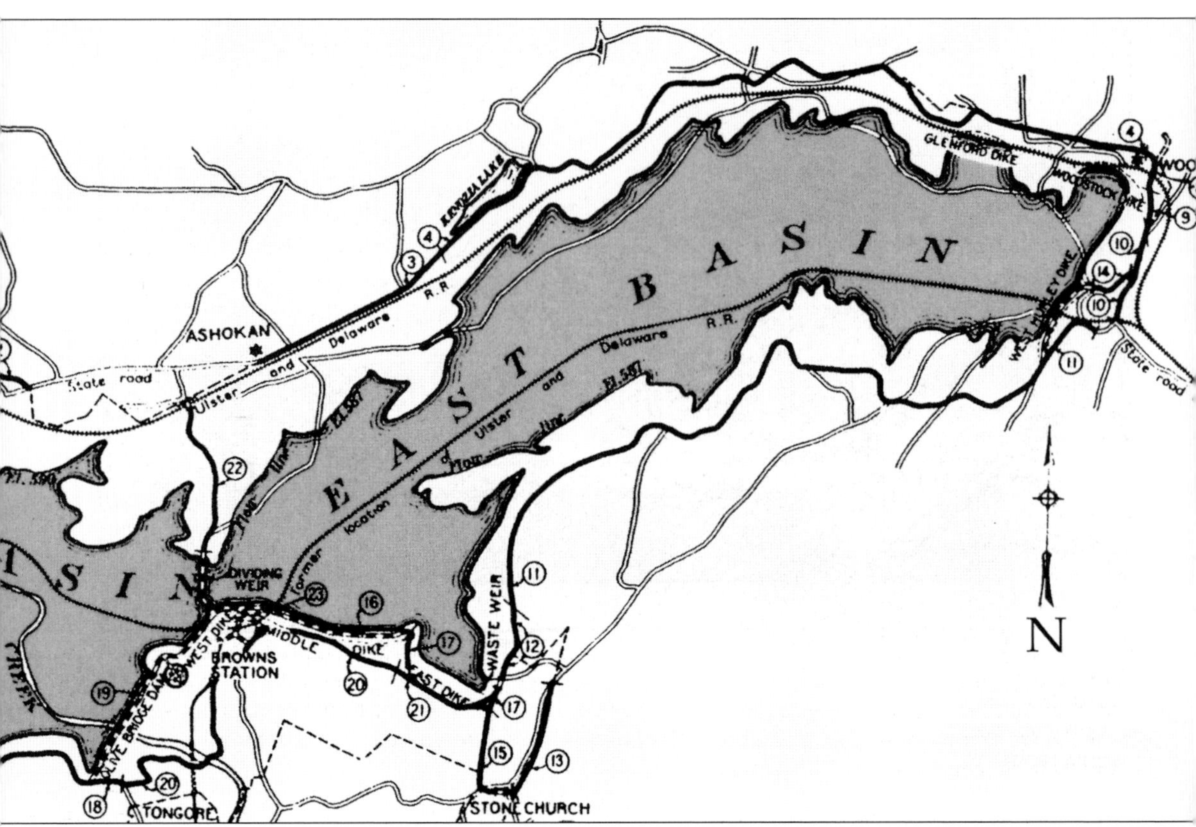

This early map identifies the dikes surrounding the East Basin. Without them, there would be no way to contain the water needed to maintain a reserve. The West Basin needed no dike other than the dividing weir and the short West Dike ending above the Aerator Basin. The Glenford and Woodstock Dikes were also part of the U&D Railroad when it was moved out of the basin.

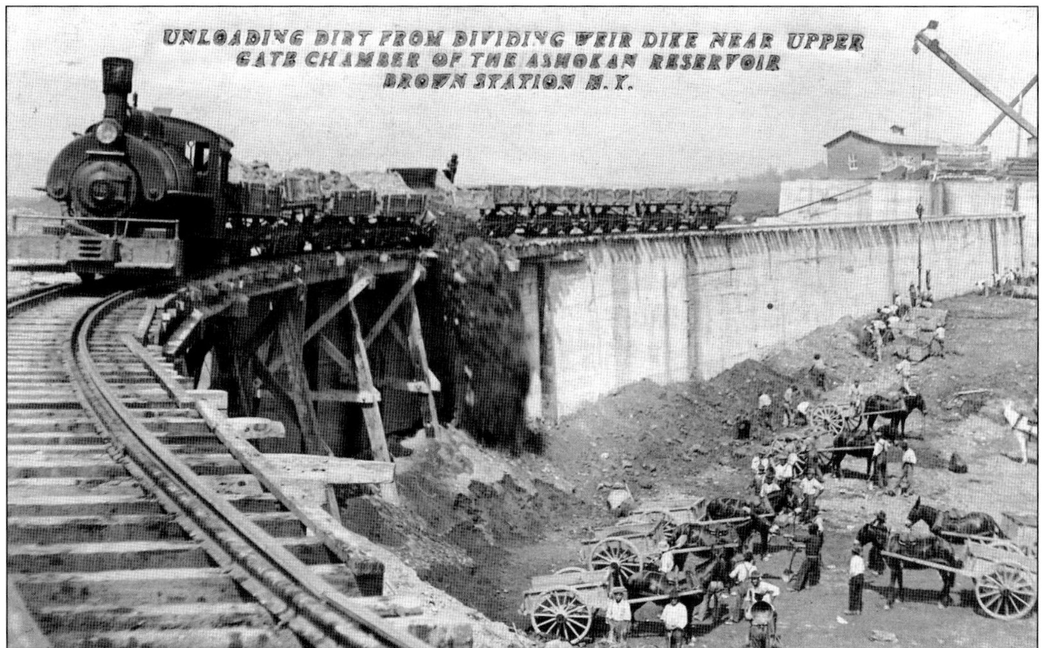

This work train is bringing fill to bank against the end of the dividing weir dike where it meets the Middle Dike, above the aerators. Workers are moving the fill away from the wall outward to make a dense barrier to the core wall. All stones in the fill are raked to the outer edges of the dike to permit the soil to be compacted.

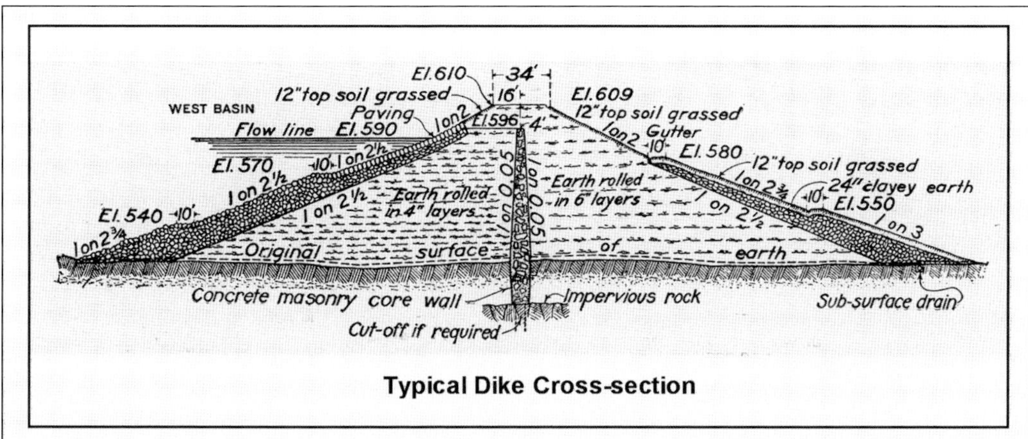

This typical dike cross-section is used on all dikes and dam wings. Slopes are approximately 30 degrees, built to 610 feet elevation, with an earthen surface 34 feet wide. The core walls are cast to 590 feet elevation. Waterside earth is rolled to four inches, and outside to six inches. Notice the outside treatment, with rocks to the waterside. The outside is covered with soil for grass.

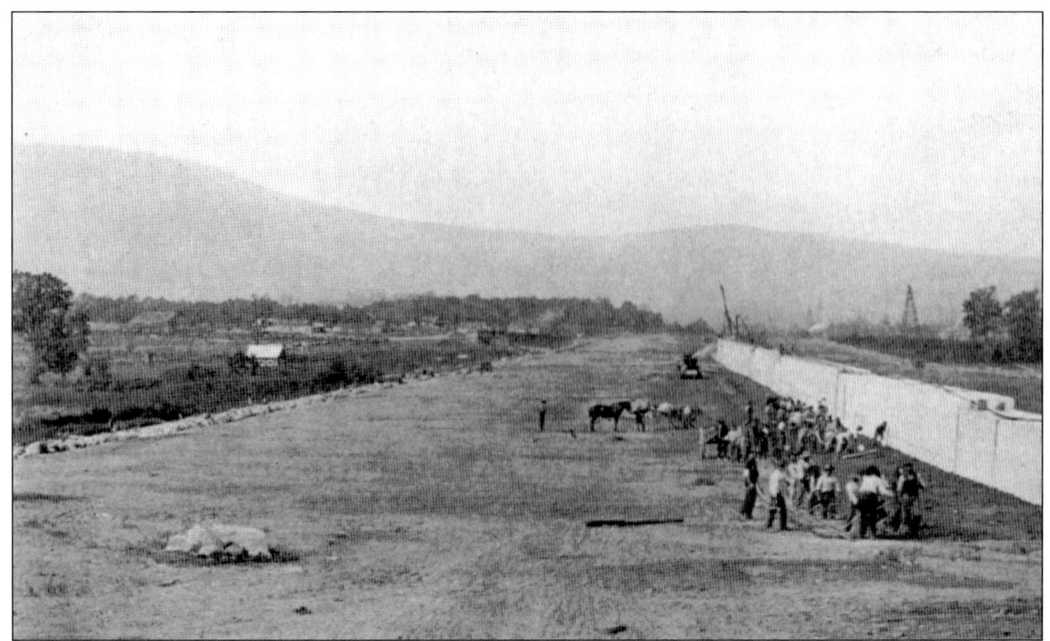

Pictured here is construction work on an early dike section. The width of the cleared area indicates the finished height. For every 2.5 feet away from the core, the finished dike is a foot higher. The object in the distance beyond the workers is a steam-powered roller. In 1985, Bob Steuding published *The Last of the Handmade Dams: The Story of the Ashokan Reservoir.* This image shows what "handmade" meant.

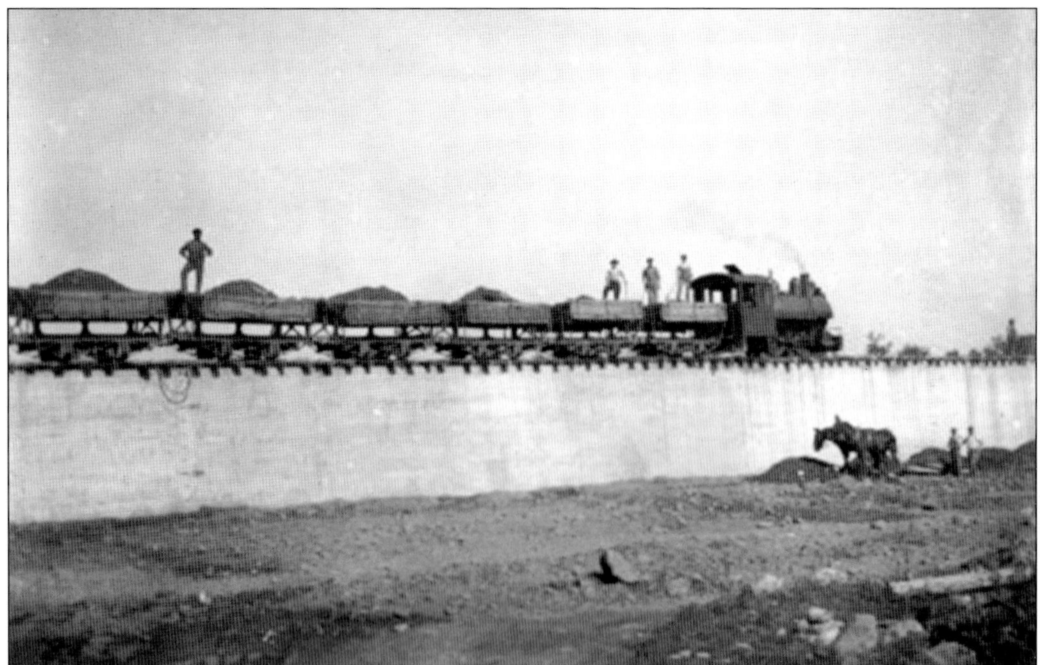

A string of specially made cars carries earth fill for both sides of the core wall. These cars are capable of being manually dumped to either side. Depending on the distance away from the wall, earth is carried by wagon or hand to where it is needed. Here, a horse-drawn sledge is used.

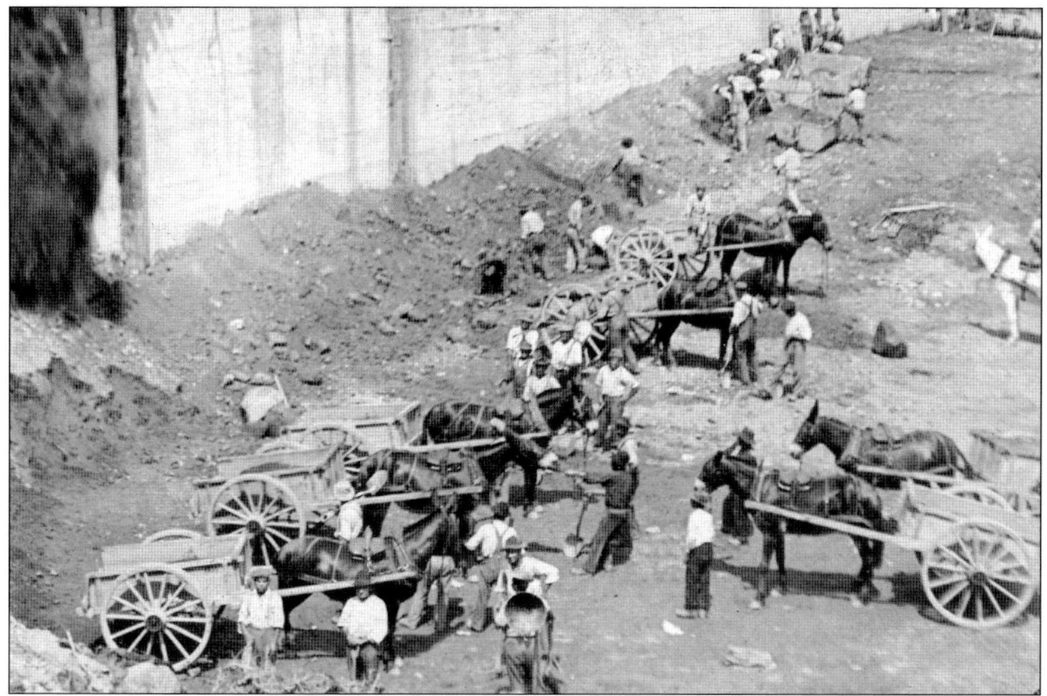

This detail of the image on page 61 shows the typical dump truck of the early 1900s. Earth was shoveled into the wagons, and the mule drivers (mostly from the South) would drive the material where it was needed. The wagon tilts backward to dump the load and then returns for another. Beyond the wagons, workers load rocks in skips to be carried away by the cableway.

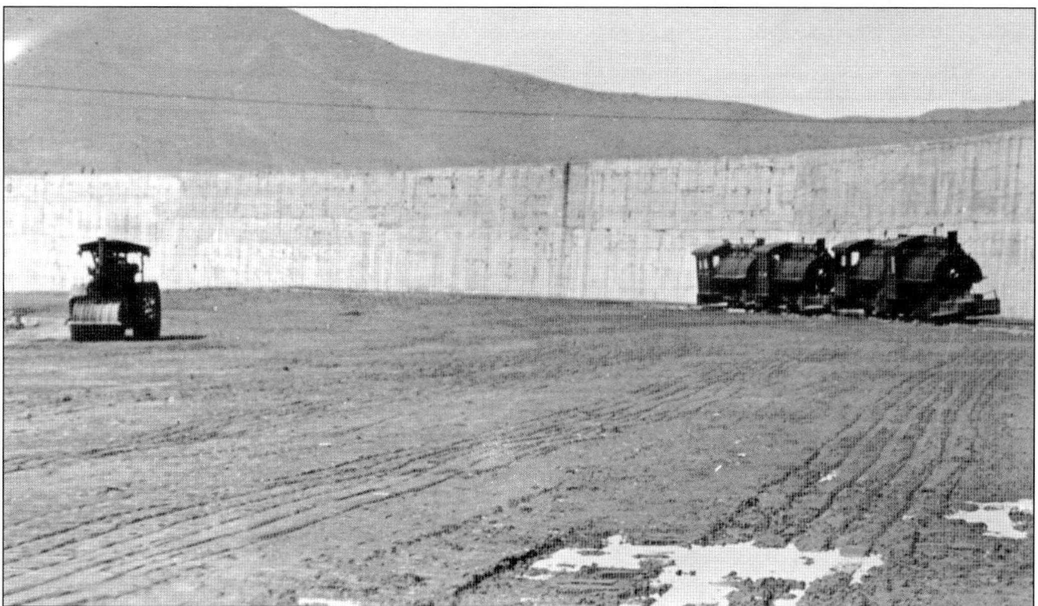

A single sheep's foot roller is compacting earth near the West Hurley Dike. High Peak is in the background to the west. This roller had inch-long, two-inch-diameter flat bottom pegs surrounding the roller drum that applies more compacting force. This type of roller was used to obtain greater compaction than typical smooth surface drums.

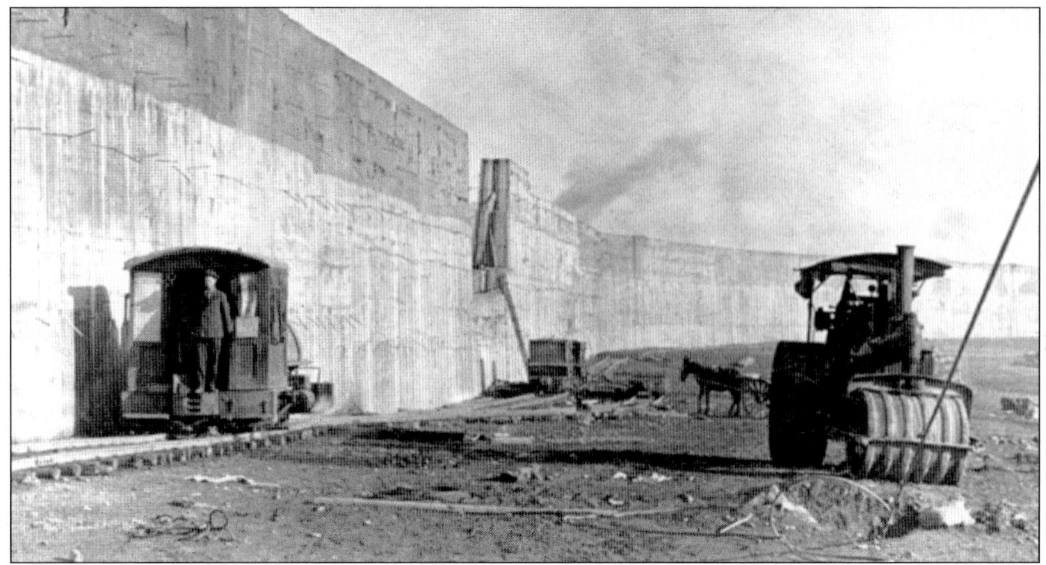

A cableway is towing another sheep's foot roller. Rollers use steam power to move, and cableways operating over 2,000 feet provided an alternative to steam. The image also shows the connection method of two completed walls. A short wall section is cast and keyed into the previously made sections. When the embankment is sufficiently built up, the upper section is poured. Narrow-gauge locomotives delivered material to the new connecting joint.

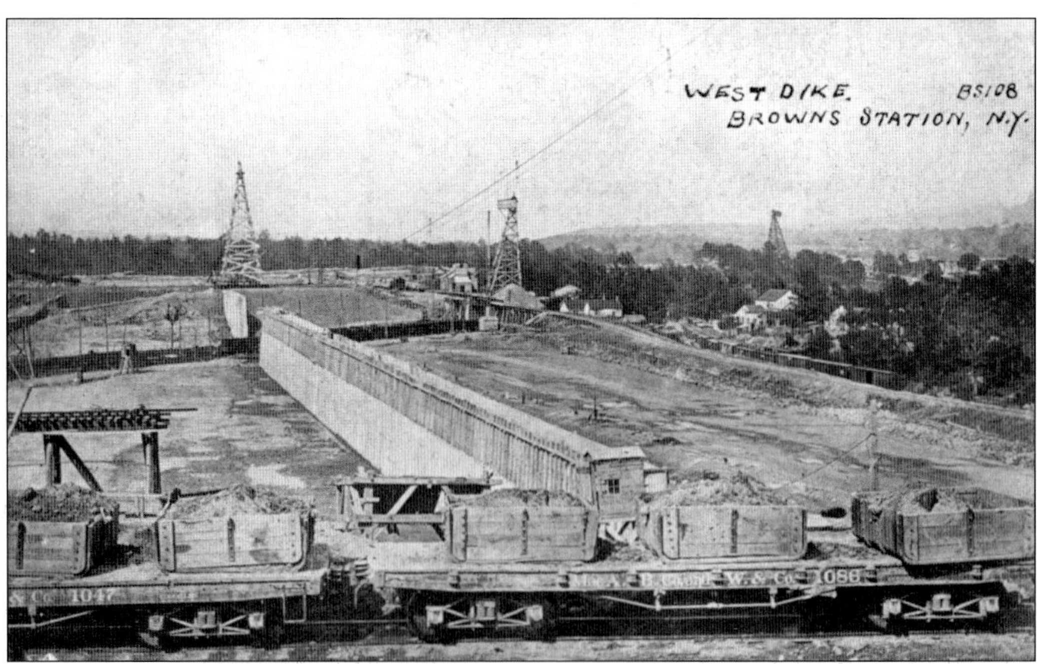

This postcard of the West Dike shows an intermediate set of forms in place on the core wall. Forward of the cableway tower, embankment material was brought to the working height of the core wall. Cableway towers could be moved laterally across the embankment area. The skips on flatcars in the foreground carried embankment fill, and much more was needed.

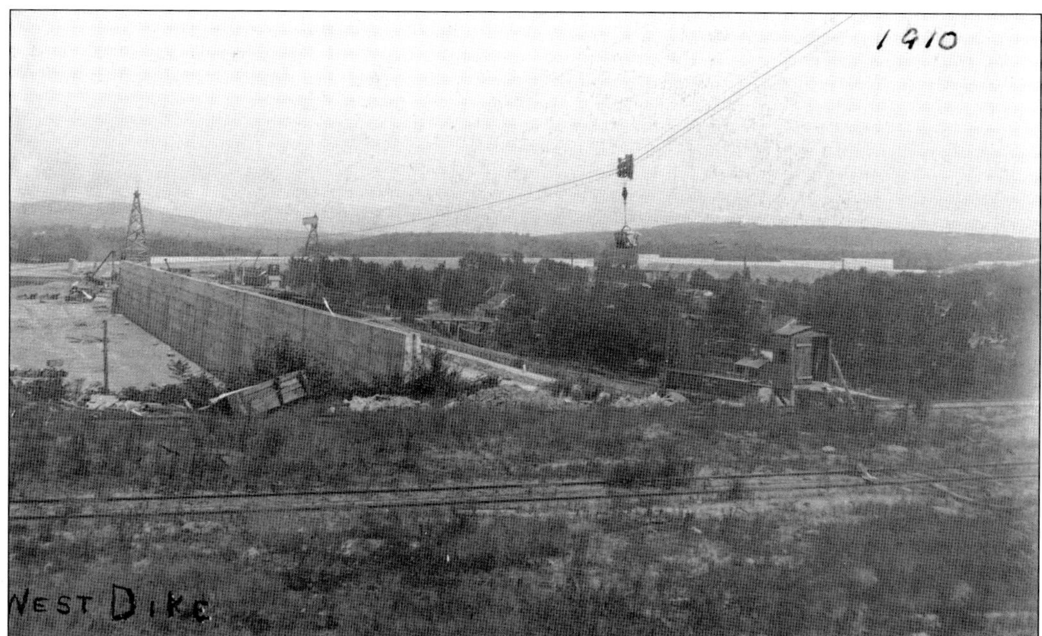

This 1910 view shows the West Dike ending at the U&D opening in front of the cableway tower. It is clear that more dike height was needed. Left of the cableway tower is the U&D portal covering. Embankment material would continue to be added over the next year or two. The Middle Dike is the white line in the distance; when completed, it became part of the perimeter roadway.

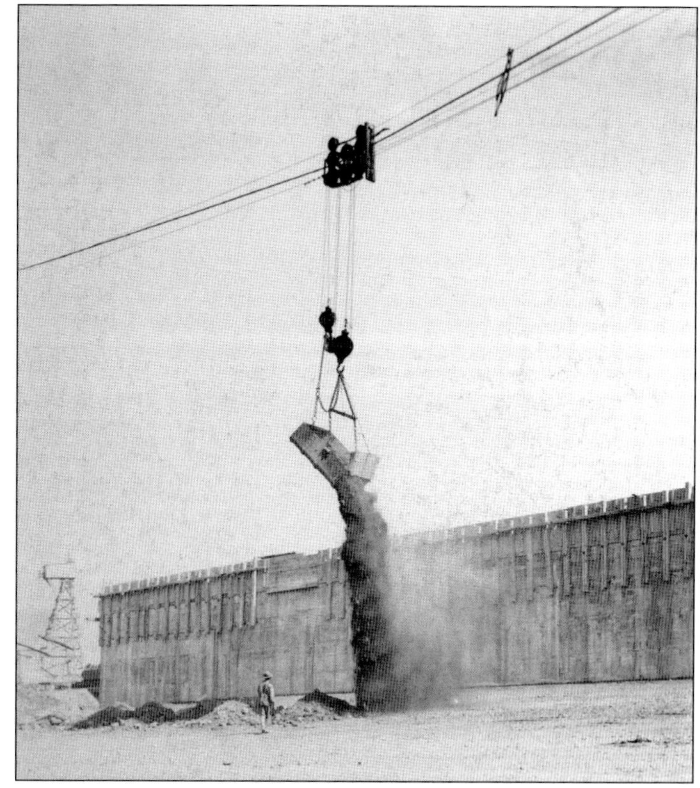
This view shows the efficiency of the cableways in moving materials where they are needed. An operator in the tower, aided by a groundworker to attach the skip, could lift, locate, and dump the load anywhere. The carriage could travel up to a thousand feet in a minute. More detail on cableway operation can be found in chapter seven. (Courtesy of NYPL digital library.)

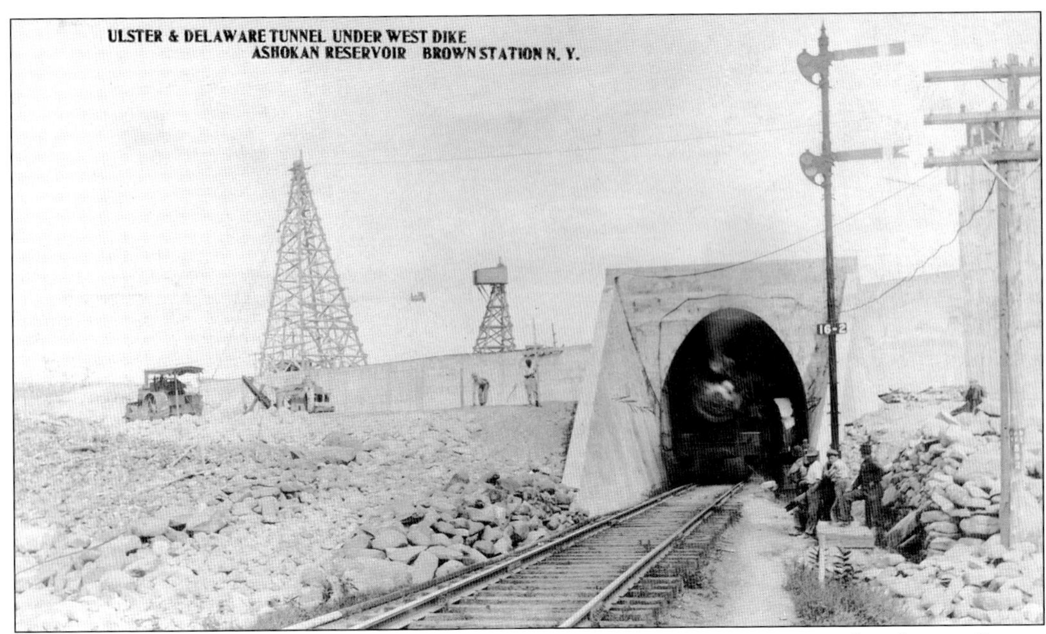

This track-level view shows the U&D tunnel with workers waiting as a train heads into Brown Station, behind the photographer. The semaphore signal arms are configured to signal "stop" for any traffic moving in the opposite direction. The signal would drop after the train cleared the station. The U&D only ran one train east or west, crossing farther west. Contractor work trains sometimes used the mainline. The number 16-2 on the semaphore pole is a mile marker from Kingston.

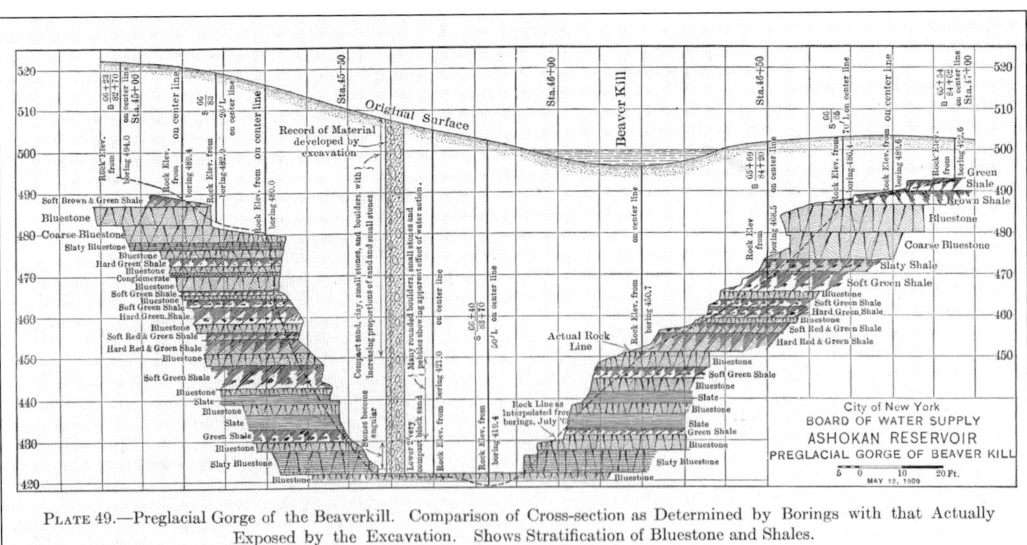

PLATE 49.—Preglacial Gorge of the Beaverkill. Comparison of Cross-section as Determined by Borings with that Actually Exposed by the Excavation. Shows Stratification of Bluestone and Shales.

This Beaverkill cross-section shows a problem contractors had building the Middle Dike center section. During earth removal in 1908, a gorge 80 feet deep with about a 150-foot top width was discovered. Excavation continued into the next year, when the 40-foot-wide bottom was found. Thousands of years of compression since the glaciers melted had made the earth so hard that removal by jackhammers was necessary.

Here, the foundation for the Beaverkill trench is underway. The sidewall of the trench can be seen below the derrick. A conduit, similar to the main dam, had to be built under this wide foundation to accommodate streamflow. Closing was similar to that in the main dam. The area in this view became the reservoir bottom. Material removed from the gorge was stockpiled for embankment construction.

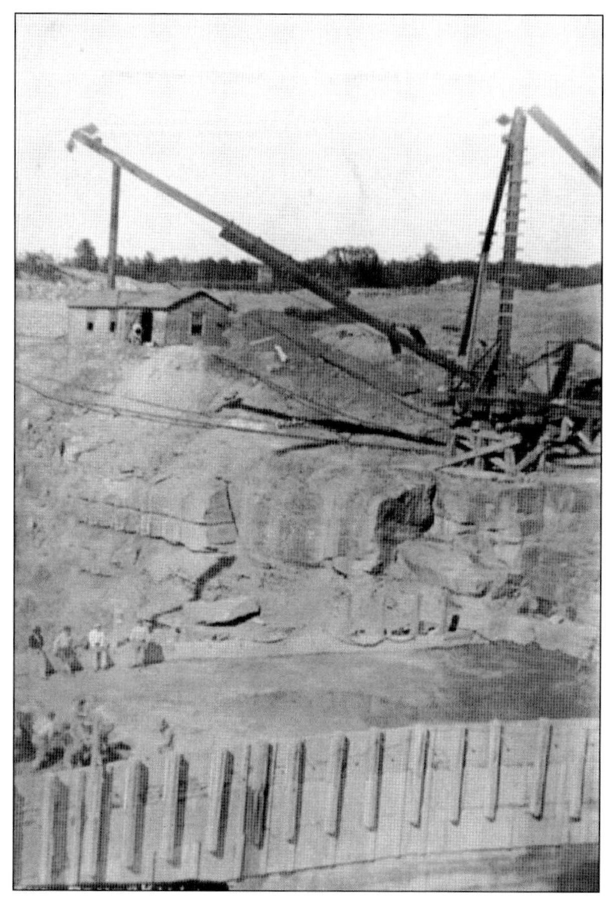

The Middle Dike is being built. The Beaverkill gorge cavity is visible in the foreground. Construction progresses on this wide section of the core wall. Estimating from the width of the dike foundation and the size of the cleared area, there is significantly more height to be added. Embankment earth is stockpiled on both sides of the tower. The next section of the dike behind the tower has not been started.

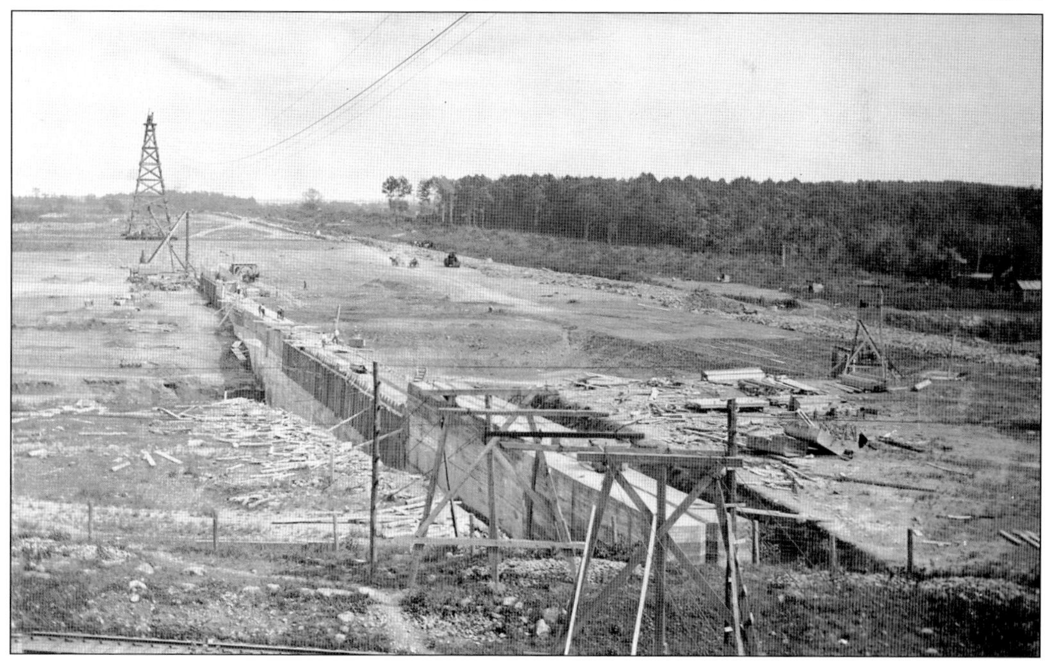

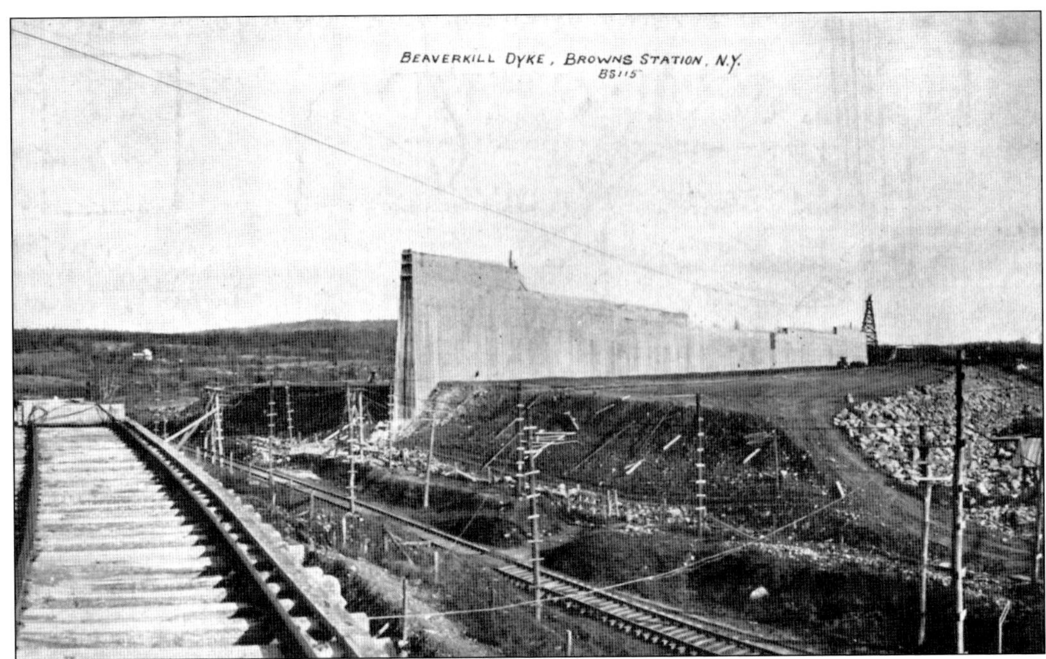

This postcard shows the Middle Dike, at the west end of the center section, built next to the U&D track to the West Hurley area. A tunnel through the dike would be built to allow for continuous rail service. The Woodstock and Glenford Dikes were used for the railroad relocation. At right are the rocks moved out of the embankment to allow for proper compression.

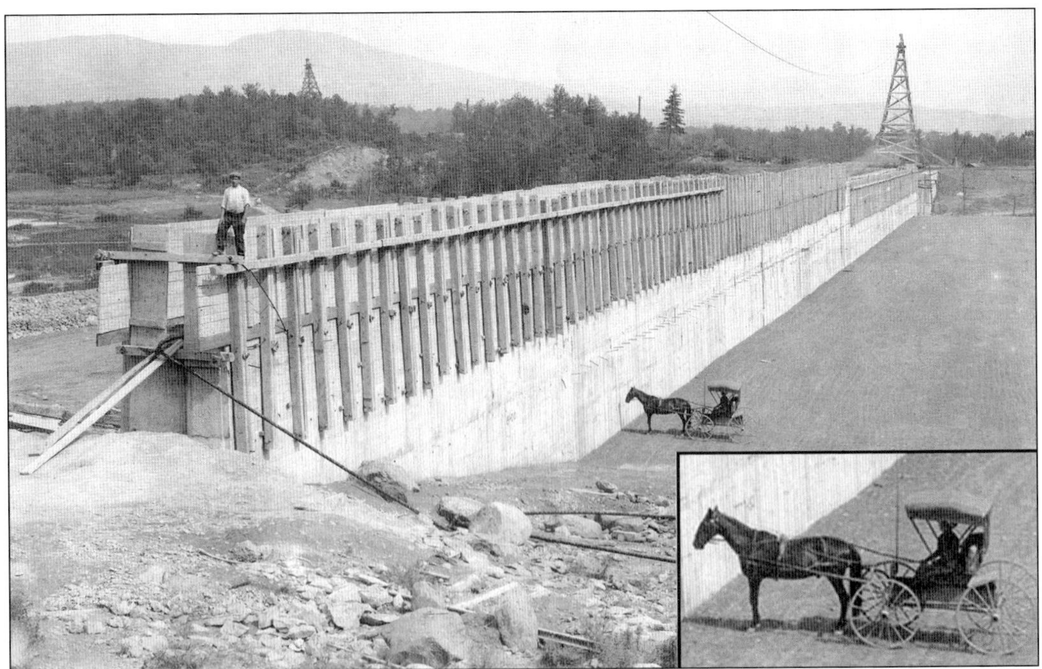

This dike section image was captured on a weekend, so no work is underway. Note the worker on the dike showing off the project and a woman in the buggy.

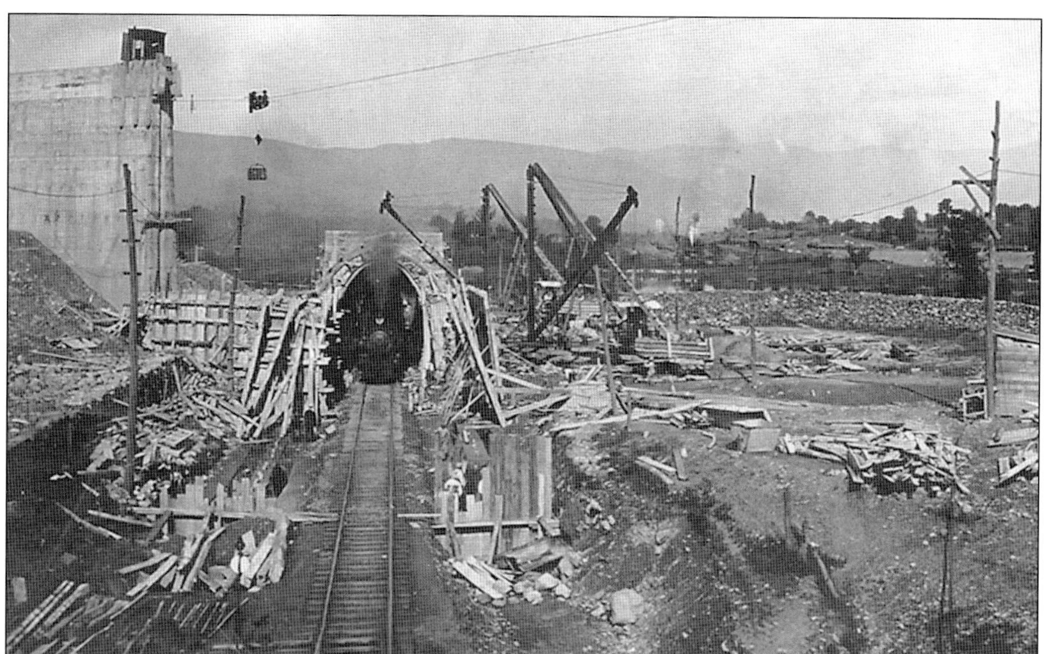

The Middle Dike center section, shown from the opposite (water) side, is being created with the U&D tunnel and a connection to the west section. This is probably a late 1911 image, as the center is at full height with the railroad operating. Earth removal had started on the west section. A cableway operators' tower can be seen behind the core wall.

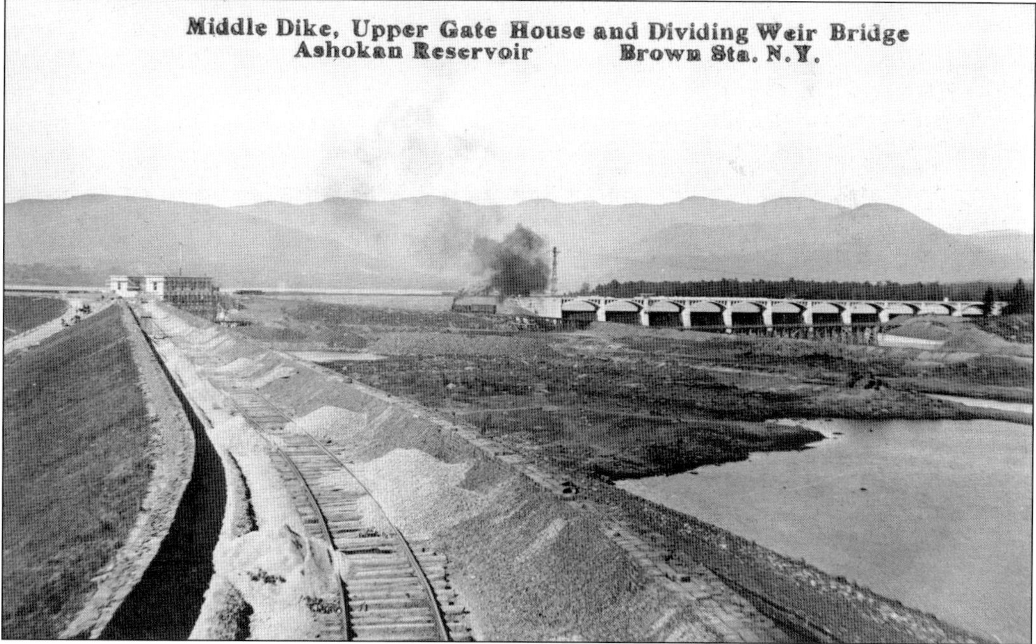

This photograph taken after East Basin water storage was established in 1915 looks across the Middle Dike at the upper gatehouses, the sluice gates (behind the locomotive smoke), and most of the arches of the nearly completed dividing weir. Narrow-gauge tracks were still in use on the Middle Dike to carry bluestone pieces to the dikes for paving.

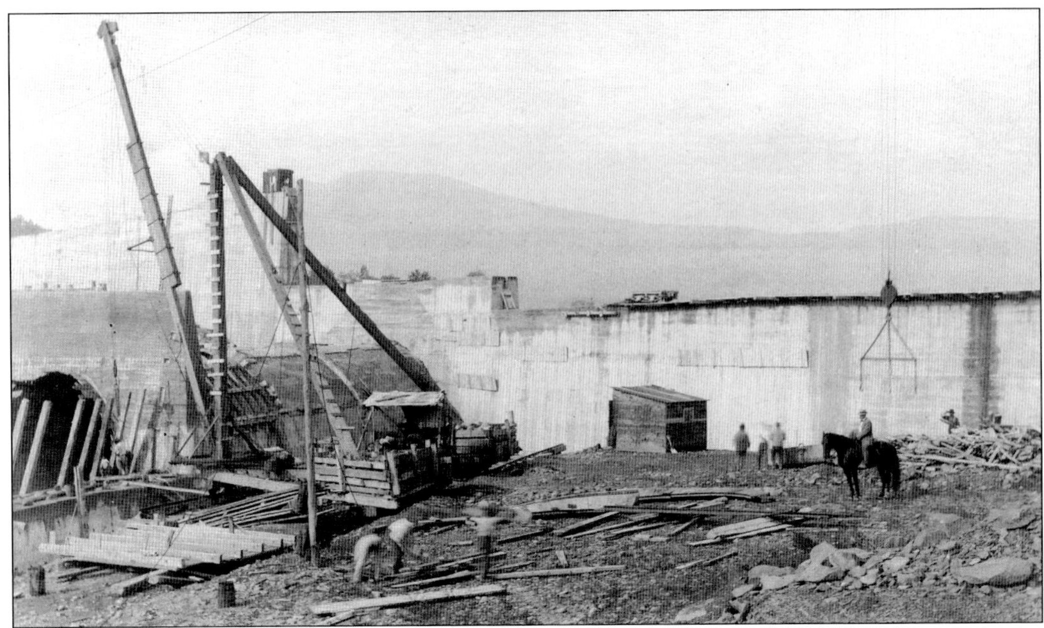

The Middle Dike's west end (see page 64) has been joined at the center section. The tunnel has been elongated to protect train operations as the waterside embankment increases in height. Forms for masonry over the tunnel are being installed. The complete Middle Dike, 11,600 feet long, closed off all East Basin lowland. Originally a roadway, it is now a popular paved walkway worth a visit.

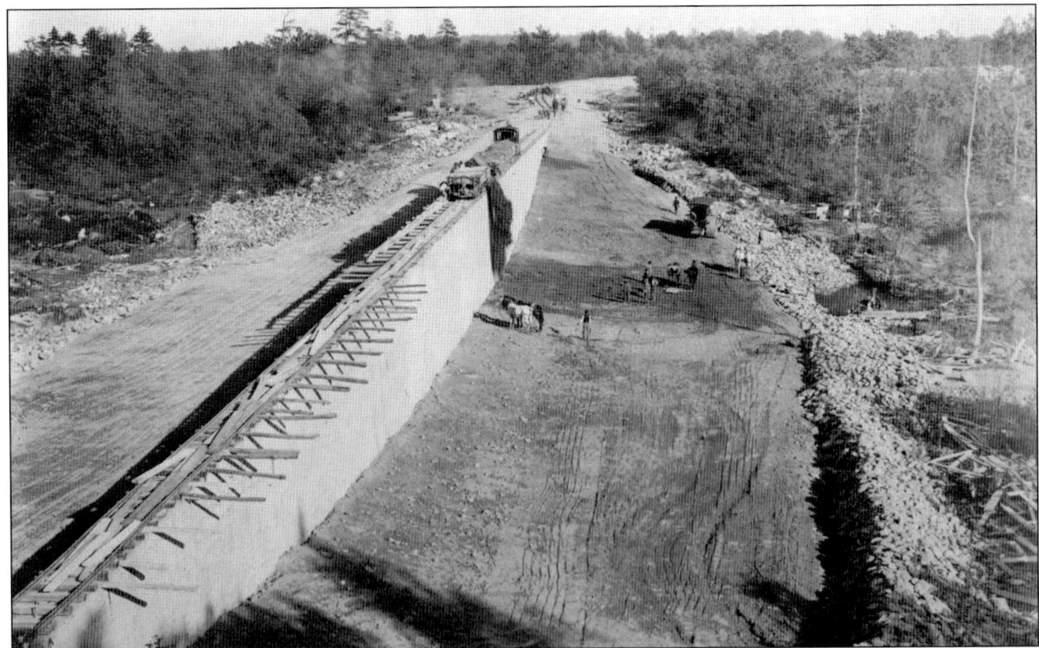

This view looks east along the Woodstock Dike from the basin's northeast corner. Embankment material was delivered by narrow-gauge railway from pits a half-mile away. Horse-drawn slip scrapers were used to distribute fill prior to compacting. Because of the distance to the main camp, workers had a small camp near here, with buildings similar to the main camp at the dam.

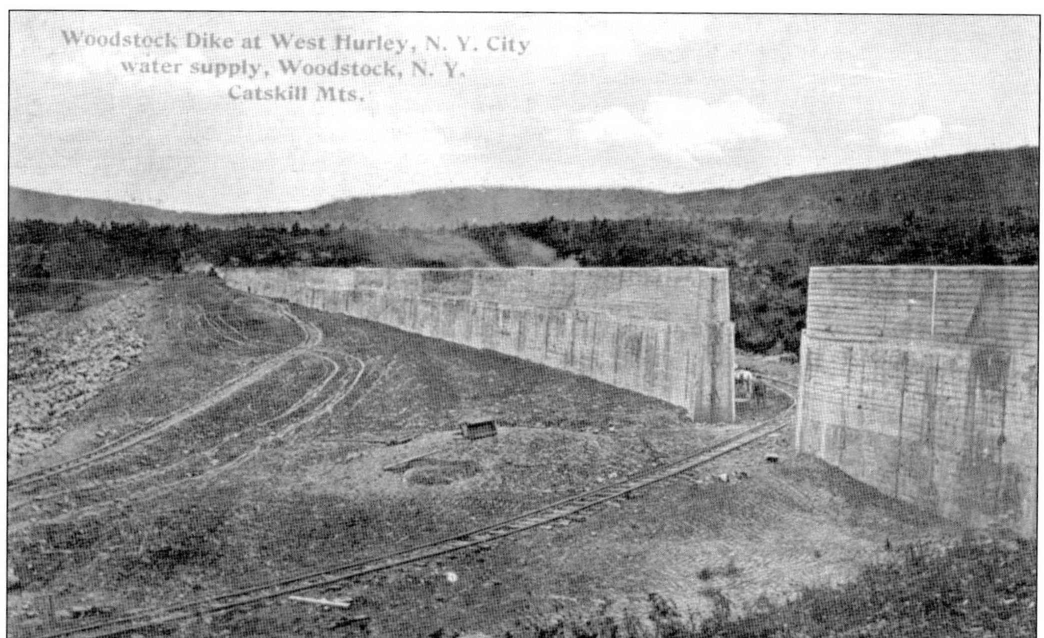

Woodstock Dike is at full height and shown in two sections, with a shallow angle between them. The opening provides access for embankment material plus a means of joining the two sections. White states that during 1910, 290,000 cubic yards of embankment was placed on this, Glenford, and West Hurley Dikes. For comparison, a large bag of garden soil is two cubic feet. One cubic yard would be a little more than eight of those.

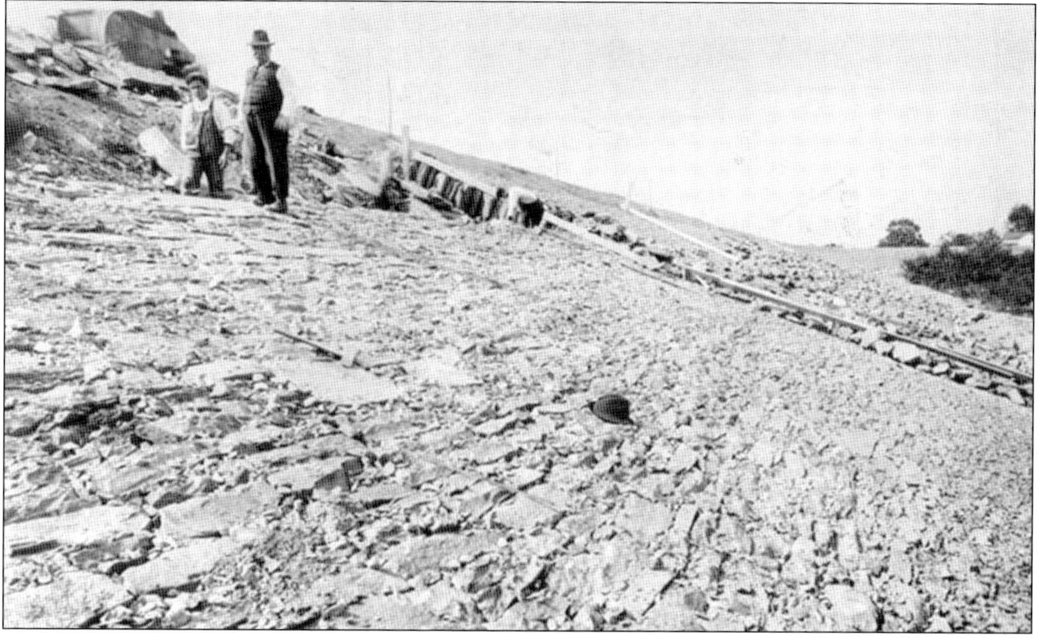

This view shows the laborious job of paving the interior of the dikes' inner walls. Large pieces of stone of varying thickness were stood on end from the beginning of the slope to 10 feet above 587 feet elevation—full pool for the East Basin. All dikes, including the waterside of the dam's north and south wings, were paved in this manner. (Courtesy of NYPL digital library.)

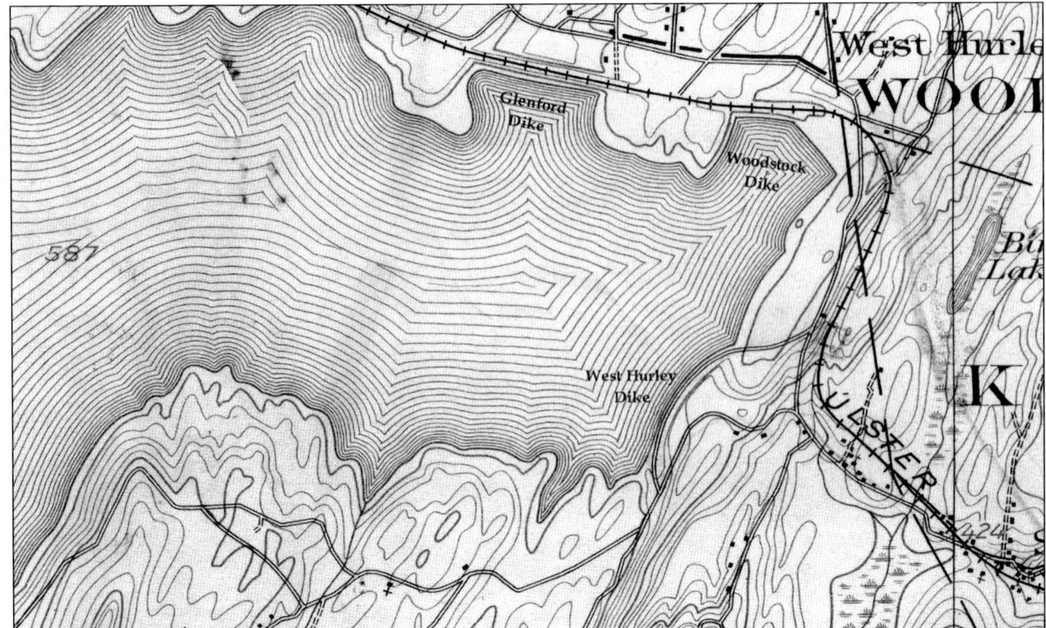

This detail from the Rosendale topographical map, updated in 1912, depicts the relocated railroad along the north shore of the reservoir. The Woodstock Dike, west section, and the full 1,825 feet of the Glenford Dike carry the new roadbed. A "Farewell" U&D train ran on September 28, 1976, after 105 years of operation. Since opening in 2019, this section's right-of-way has been one of the more popular rail trails in the region.

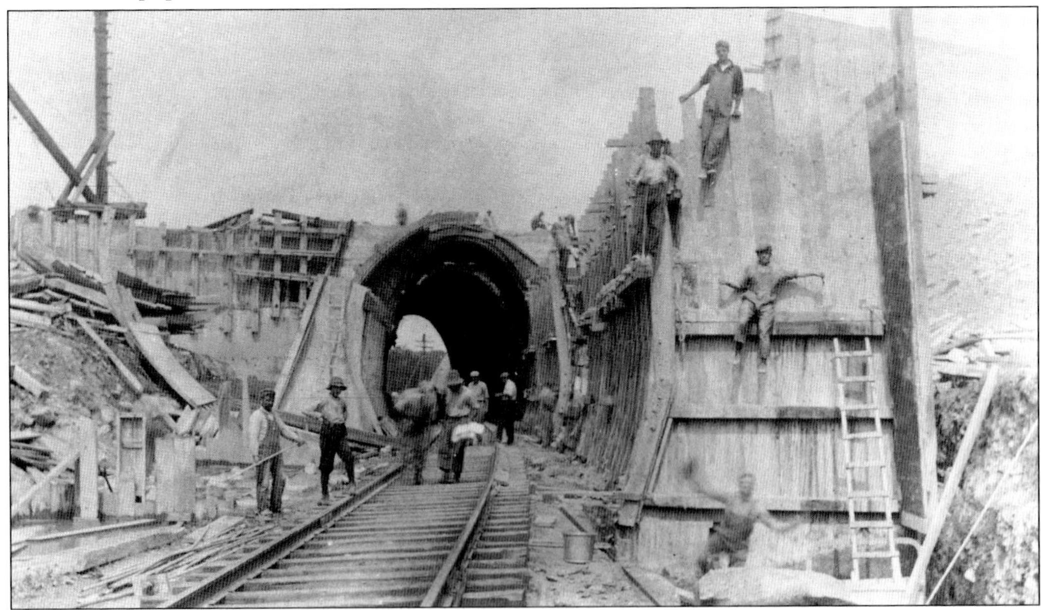

The West Hurley Dike tunnel covers the U&D tracks after exiting the basin and turning to Kingston. This dike was built using the same method as the others: core wall and embankment. Photographer William Longyear lived on Clinton Avenue, uphill from the U&D station, according to the 1907 Kingston Directory. He took the train to the site and took pictures all day. In the evenings, he printed postcards. He could step out his lower door and walk to the station.

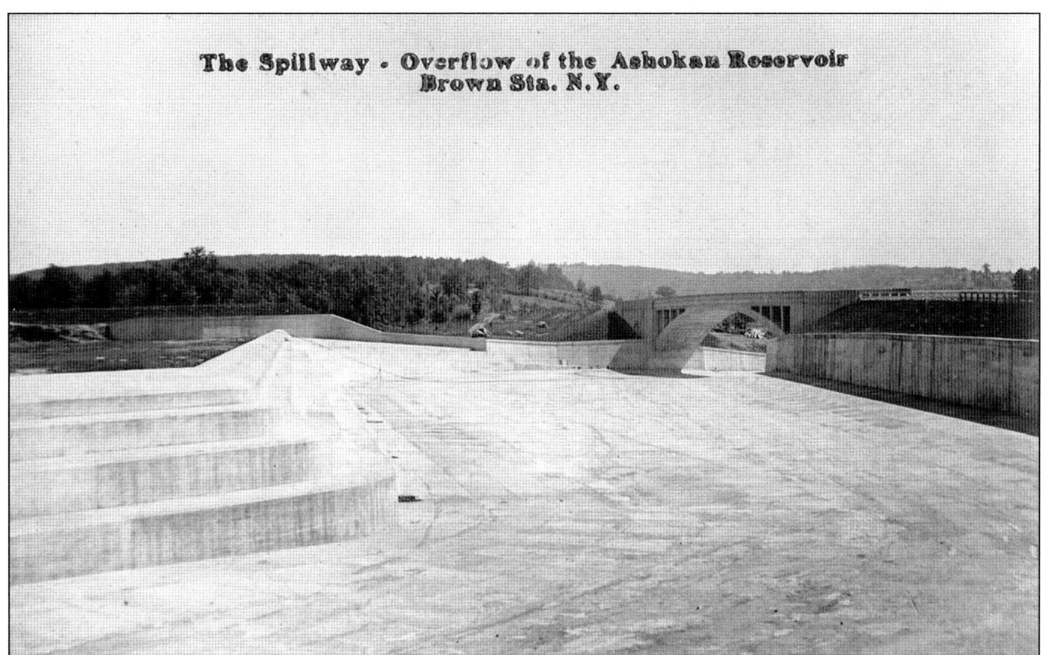

The waste weir, or "spillway" or "overflow," is about 1,000 feet wide and made of cyclopean concrete, including a deep cutoff wall under the masonry work. As with all reservoir visual parts, the design was more than utilitarian. The weir provided a gentle overflow into the specially designed collection basin, then under a bridge into a natural canyon, and into the lower Esopus Creek.

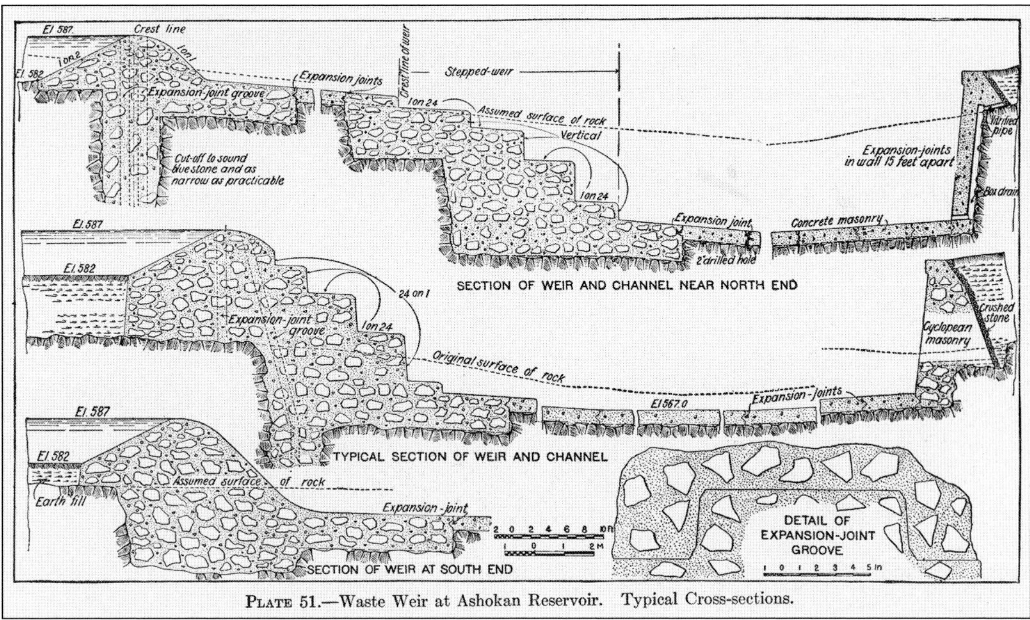

Plate 51.—Waste Weir at Ashokan Reservoir. Typical Cross-sections.

These spillway section drawings show its design at three locations. Significant rock was removed and used in the masonry. An adjacent concrete plant used a narrow-gauge railway to move prepared masonry. Expansion joint lubricant, similar to that used in the main dam, was used. The cast sluiceway floor was made using alternating blocks finished by hand. Five feet of water is retained behind the 11-foot barrier.

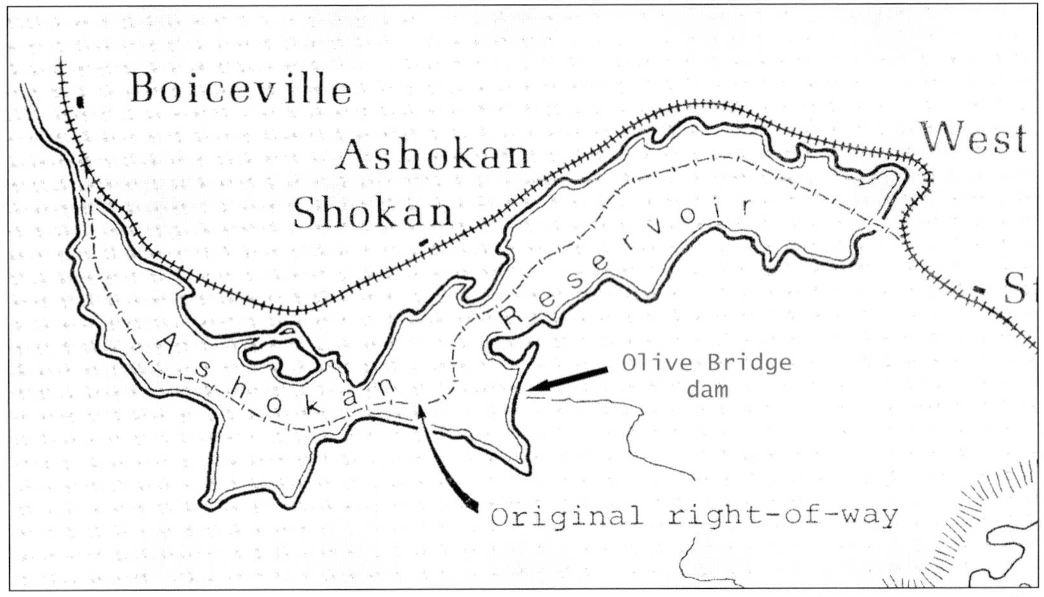

The U&D Railroad finally had to move its route outside the reservoir work area. Water in the West Basin was rising. The first train ran over the Woodstock and Glenford Dikes on June 8, 1913. This map shows the old and new routes. Brown Station was at the downward loop of track to the right of the arrow pointing out the old right-of-way.

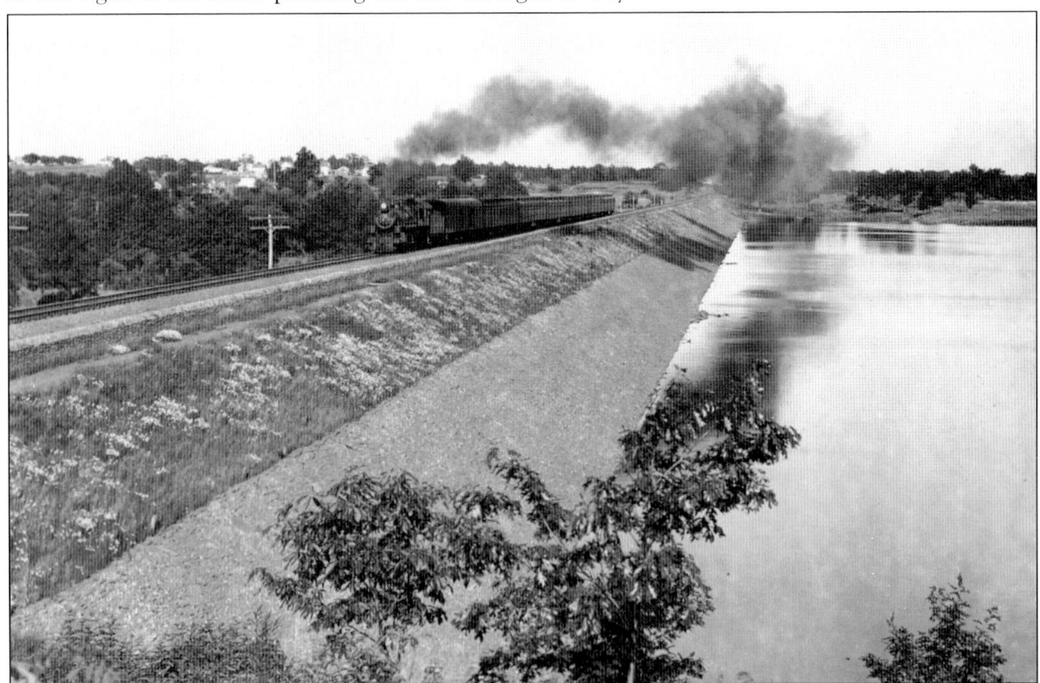

In this June 1916 photograph, the Rip Van Winkle Flyer, the U&D's premier passenger train, heads west toward Phoenicia over the new Glenford Dike. This image shows paving stabilization on the inboard side dike to prevent embankment erosion from ice and wave action. Imagine building almost six miles of this stonework. This embankment is on the east end of the new Ashokan Rail Trail. (Courtesy of NYPL digital library.)

Five

Workers on and off the Job

The workers on the project were either employees of the Board of Water Supply or people the contractors hired. Their lives were much the same as with any other job, with work and personal time. Early into construction, the workweek changed from 48 hours to 40, adding more to the cost of labor. From the beginning of construction, BWS and contractor employees worked a 10-hour day. There is no information found that discusses workers' day or lunch breaks other than a starting and quitting steam whistle. *Ashokan: Illustrated and Descriptive Account of the Main Dams and Dikes of the Ashokan Reservoir* states that the main dining hall could accommodate 200 people at a time, as a "great many men who live at home or away from camp take their lunch here and it is the busiest place on the work."

Reference books describe workers as newly arrived Italian immigrants, who in time made up half of the contractors' workforce. Austrians and Russians, along with the Italians, were about three-quarters. A New York City group, the Italian Immigrant Society, helped find them work, with some ending up at the reservoir. Society members also guided the new workers to Brown Station for employment. This group was instrumental in setting up a night school to teach English to all interested. Bob Steuding's book *The Last of the Handmade Dams* relates the positive attitude the Italian workers had toward the job and camp life. Also included on-site were approximately 150 Southern black men, some with families, brought to the job to drive mule teams. All ethnic groups lived in the camps, but in independent groups. Some local families with needed skills moved to the site, as the camp provided cottages for single families and a school building.

This chapter is presented in two groups: the jobs (though not all are discussed) and time off, which was similar to time off today, except employees did not have cars. A round-trip ticket on the train was $1.06. Weekend activities seemed to include many visitors. Local residents took great interest in the mix of nationalities on the site. They took advantage of the dining hall and restaurants until sometime in 1911, when the area was completely fenced.

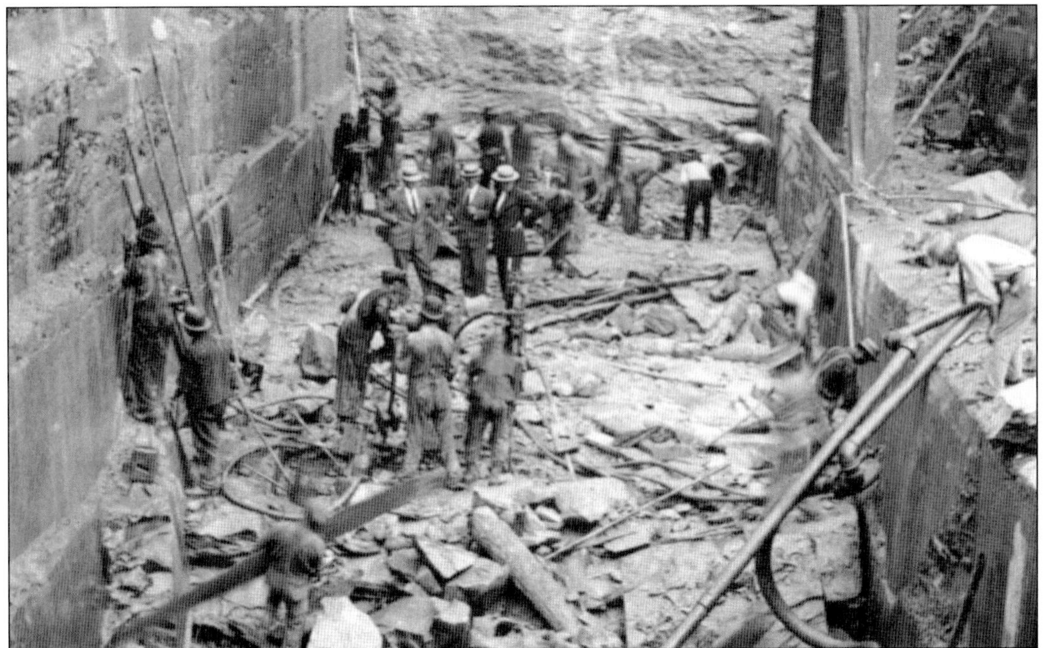

Here, workers are finishing sidewalls to remove loose rock, with others loosening rock for removal to get to a specified depth. Tubes on the right carry compressed air over a mile to run jackhammers. Explosives were not used to prevent rock fractures. The men in the straw hats may be BWS inspectors. When finished, the trench was cleaned of all loose material prior to masonry work.

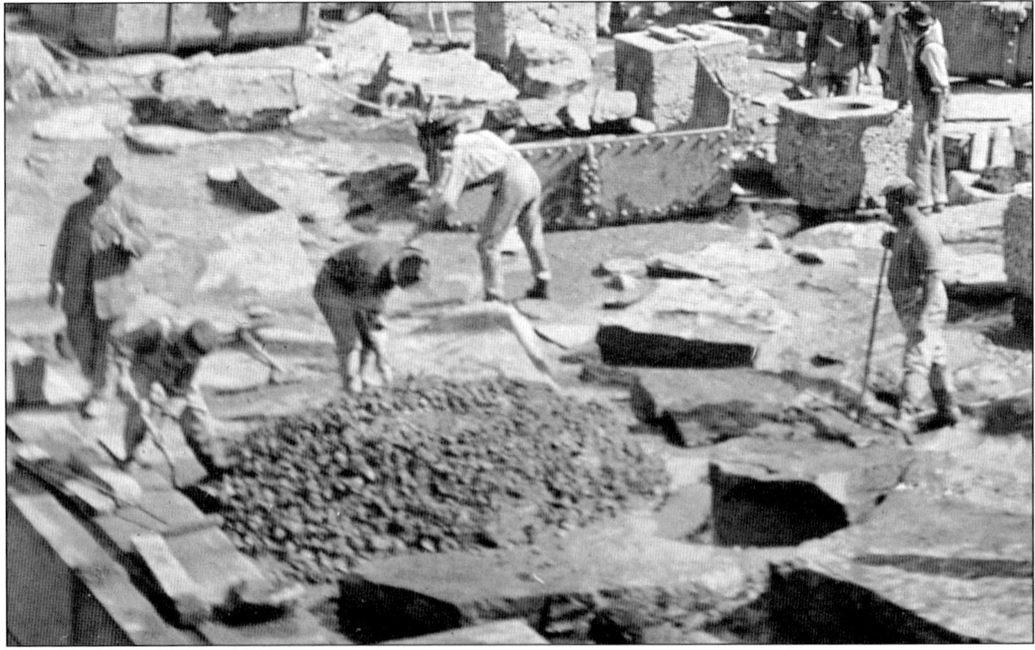

A new load of concrete is being shoveled to expose the upper third of this rectangular rock. Some masonry is moved left against the precast block, which also is a form for additional concrete. The exposed cemented rock acted as a lock when additional concrete was added. Most of the workers spent their days doing physical labor.

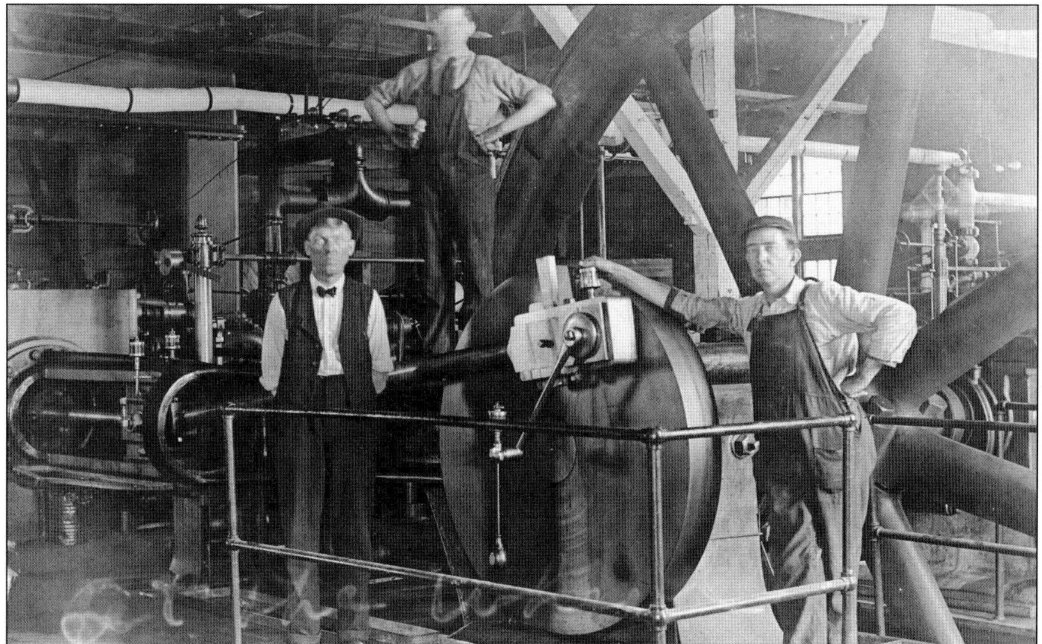

The workers who operated the powerhouse had to stand still as this photograph was taken. Note that the upper worker moved his face during the exposure. This coal-fired plant generated steam for air compressors and electric generators for all site lighting. The 14-foot flywheel was delivered by the U&D railway to the site in three pieces. (Courtesy of David Turner.)

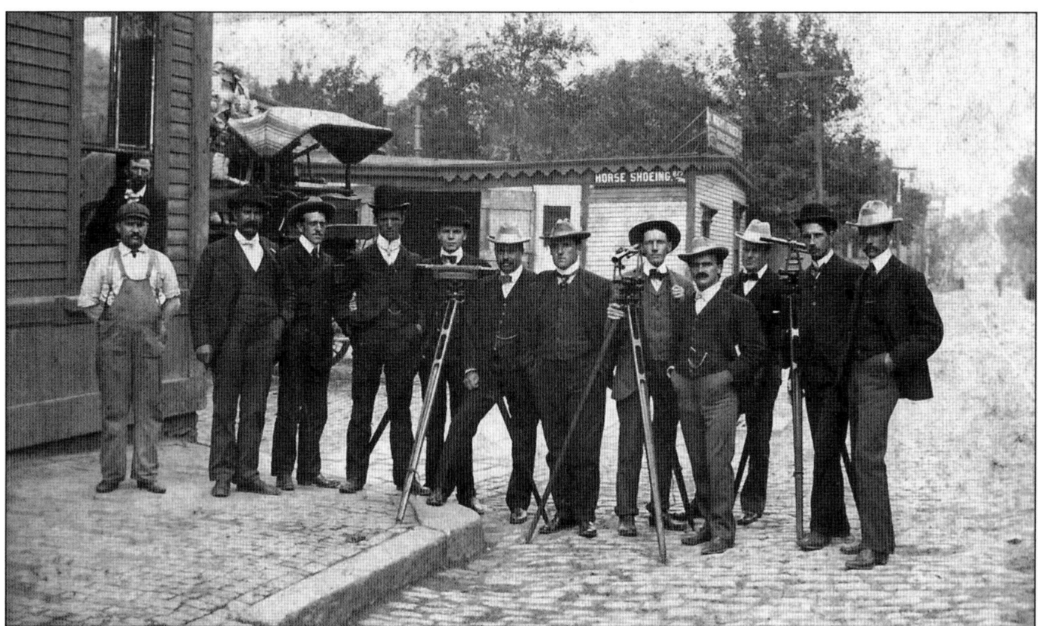

This team of BWS employees was responsible for contour mapping the entire basin within the taking line. Every property was surveyed. Their key work was to establish specific points of the whole project, such as the starting point for the dikes or where the starting corner for the Olive Bridge dam should be positioned. (Courtesy of Town of Olive.)

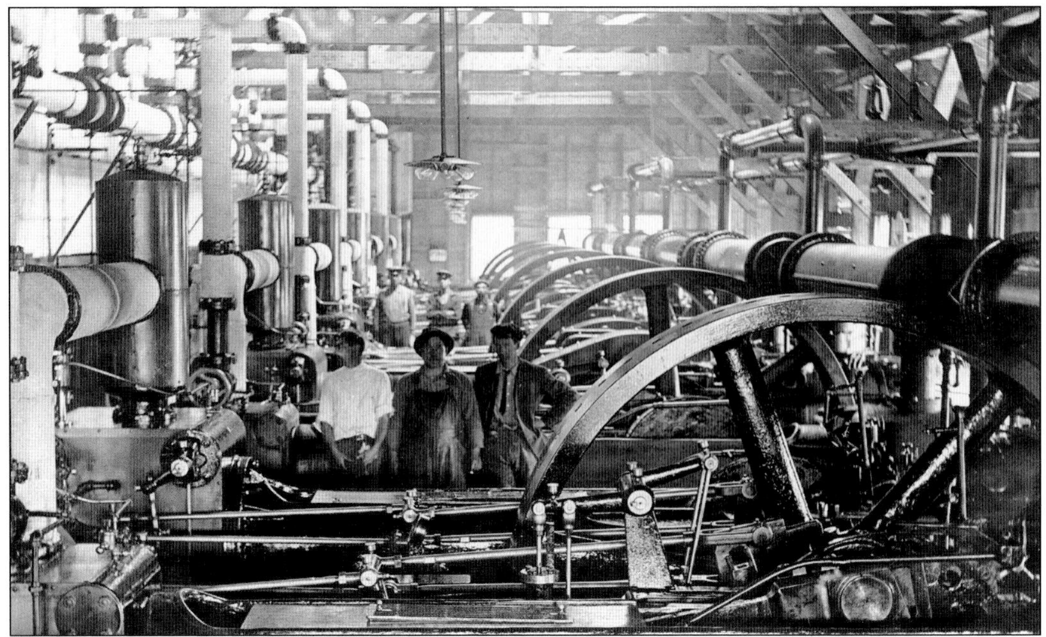

Compressed air was the prime power for almost all machines. These workers kept the four compressors operating that delivered 1,200 cubic feet per minute at 80 psi. Each machine had two flywheels. (This image shows seven.) Each unit fed its own air-line system. Compressed air was sent as far as three miles to the Yale Quarry. This area was hot in both winter and summer; as air is compressed, its temperature increases. (Courtesy of David Turner.)

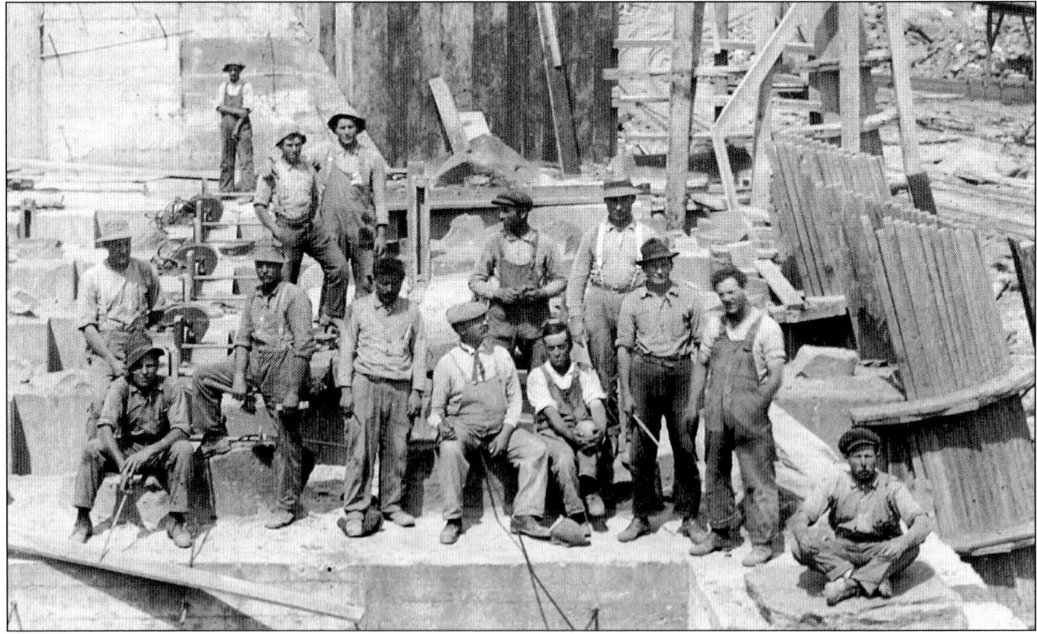

These workers appear to be a framing crew that built forms for the dividing weir masonry work. Their work moved from south to north, building three sets of forms in preparation for cement delivered by cableway. Once dried, they used derricks to move the forms to the next location north. The weir was completed by the end of 1915. (Courtesy of NYPL digital library.)

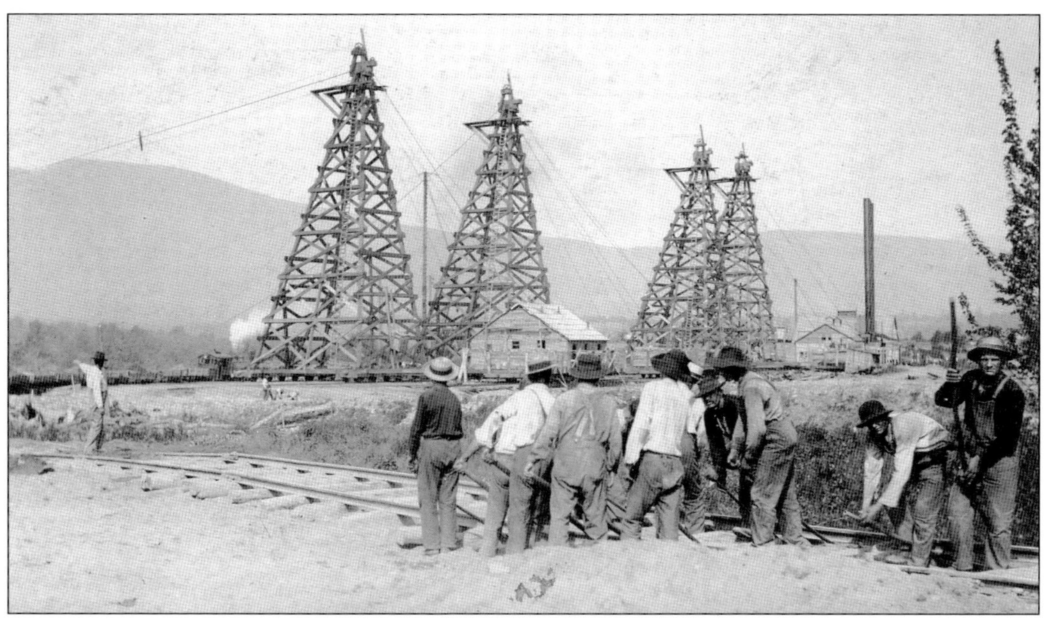

This crew is moving a standard-gauge track from a depression where the single man is standing. The track appears to take a drop there. Upon a closer look, there is a low train with the locomotive in front of the cableway towers. In 1910, the contractor had 33 standard- and narrow-gauge locomotives, as the railway was the only way to efficiently move materials. The dump truck as it exists today was not invented yet.

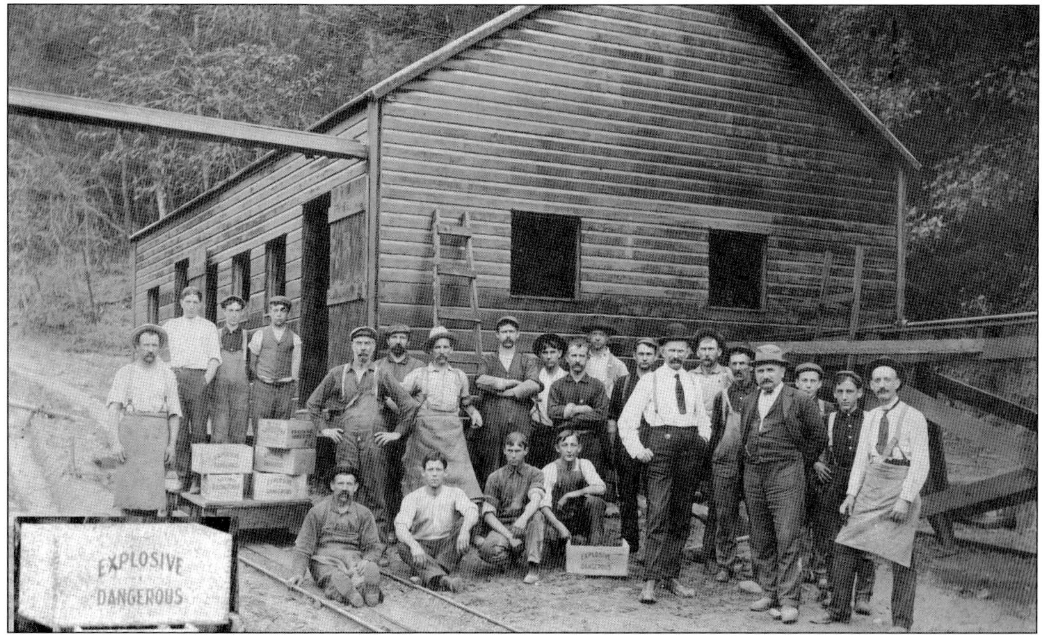

The job of these workers was unknown until the "Explosive" box was "blown up" and added as an insert. All were employed by Winston & Co. The reservoir required the use of powder in many places except the dam. Dynamite was used on the spillway and dividing weir and other places where large layers of rock had to be removed. The storage building appears to be well away from the main camp.

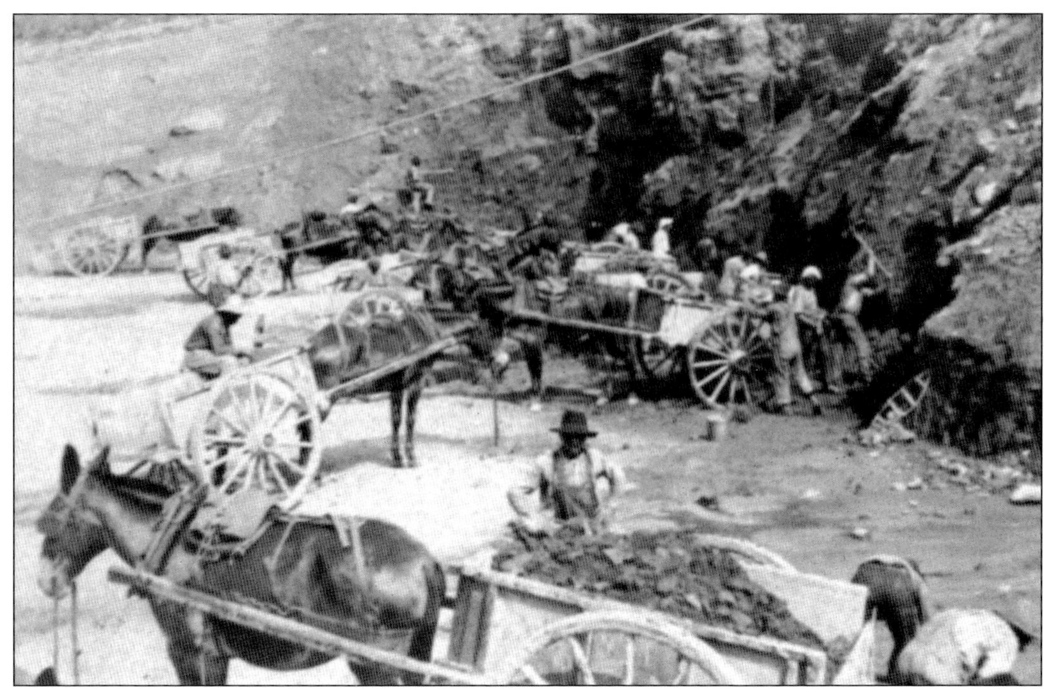

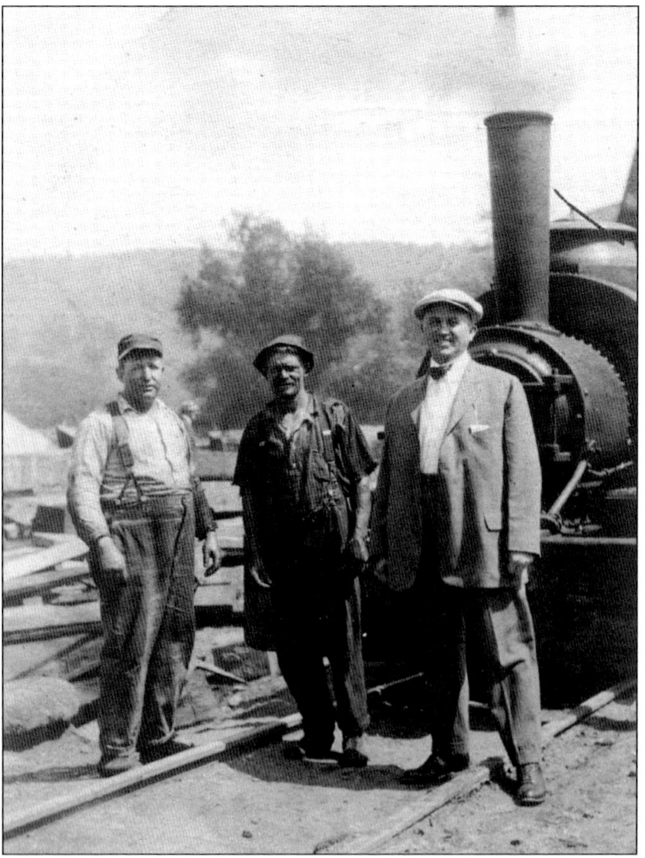

Here, mule wagon drivers assist in an earth-moving operation to supply embankment construction. Tens of thousands of years ago, when this area was a glacial lake, eroded material settled in areas of still water, providing different materials in differing locales—sometimes many feet thick, as seen here. Material was moved the old-fashioned way.

This picture shows a narrow-gauge locomotive crew. It is easy to identify jobs. Steam locomotives need a coal fire to run. They also need a driver, and the driver needs orders. Where to go for a load of material and transport it to its destination is the conductor's job. Nothing in the basin is close. It is about six miles from the dividing weir to the West Hurley Dike.

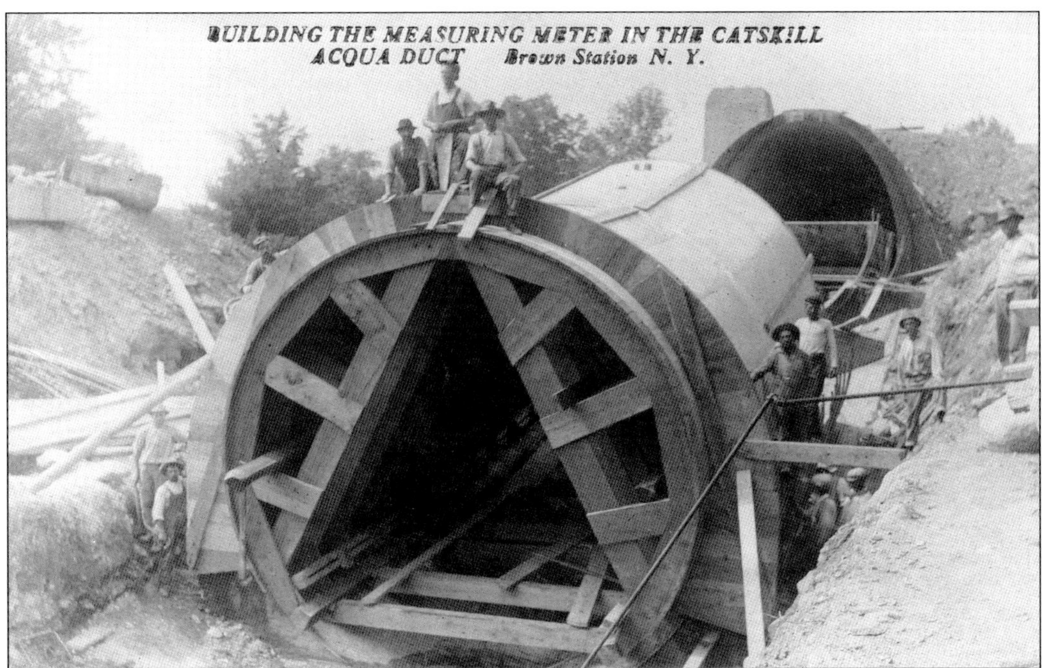

In 1910, this Venturi meter was the world's largest of its kind. The BWS needed a regular measurement of reservoir water outflow. By measuring pressure at two different aqueduct diameters (17.5 feet and 7.75 feet) and a little arithmetic using Bernoulli's principle, the flow rate could be calculated. Workers are connecting two precise diameter cast sections to the outflow of the reservoir. Carpenters are building the upper concrete form; the lower concrete is done.

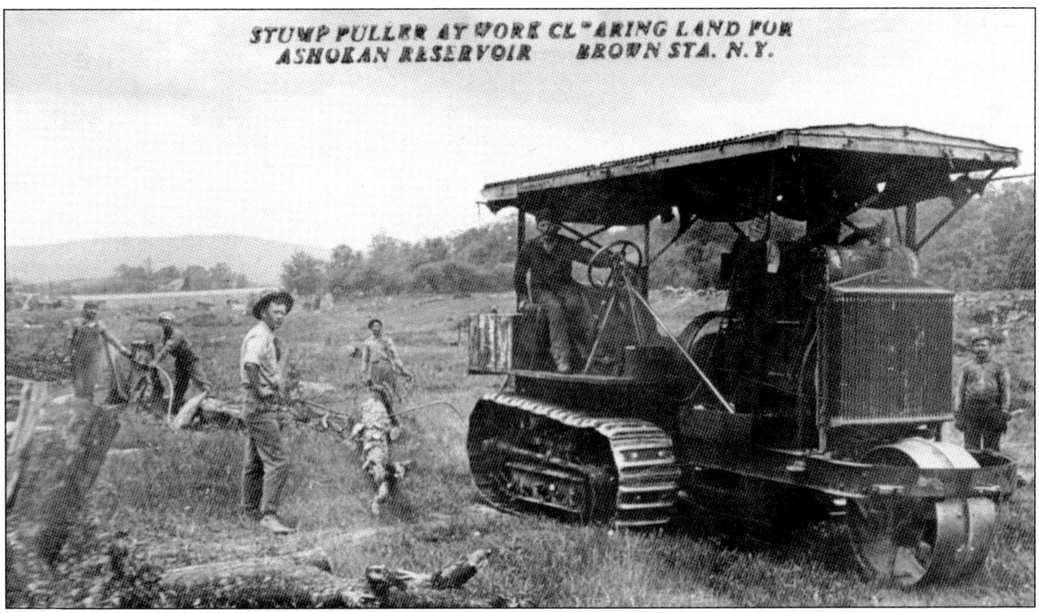

Clearing 13 square miles required some advanced help. All vegetation above an inch diameter was removed and burned. This Holt 75 model gasoline-powered tractor, first available in 1914, was the beginning of the Caterpillar Company. According to Steuding, the tractor could remove one 10-inch tree every minute, with helpers doing the cabling between the tree and tractor.

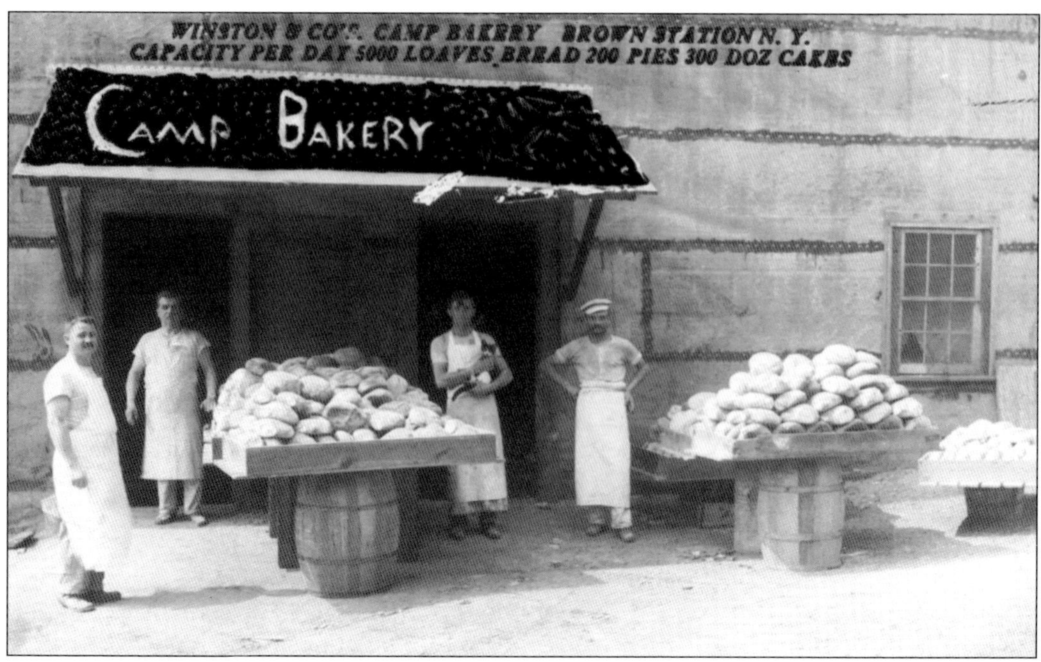

The camp bakery was near the cookhouse and dining hall and supplied baked goods for every meal. About a dozen skilled bakers worked here as part of the Winston & Co. staff. They may have moved from the Kensico Dam, where Winston was the contractor prior to the Ashokan project.

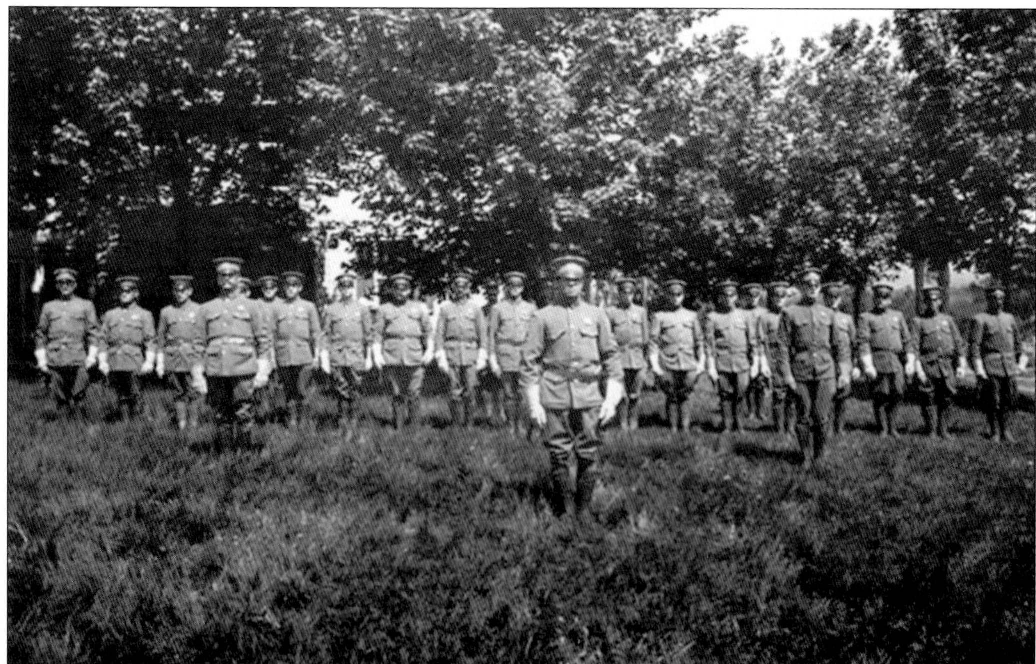

The Ashokan police battalion is shown here. Formed by BWS to maintain order between workers and local residents, a squad of 11 was assigned to Brown Station. Alcohol was not allowed on site; however, many bars were close to the camps. The force is active today and has the same authority as local or state police. (Courtesy of Town of Olive.)

This group of skilled workers probably moved here with Winston & Co. as regular employees. They were kept on to maintain the large number of horses and mules used throughout the project. Horseshoeing, harness making, and repairs were constantly required, with over 500 animals on site. (Courtesy of Town of Olive.)

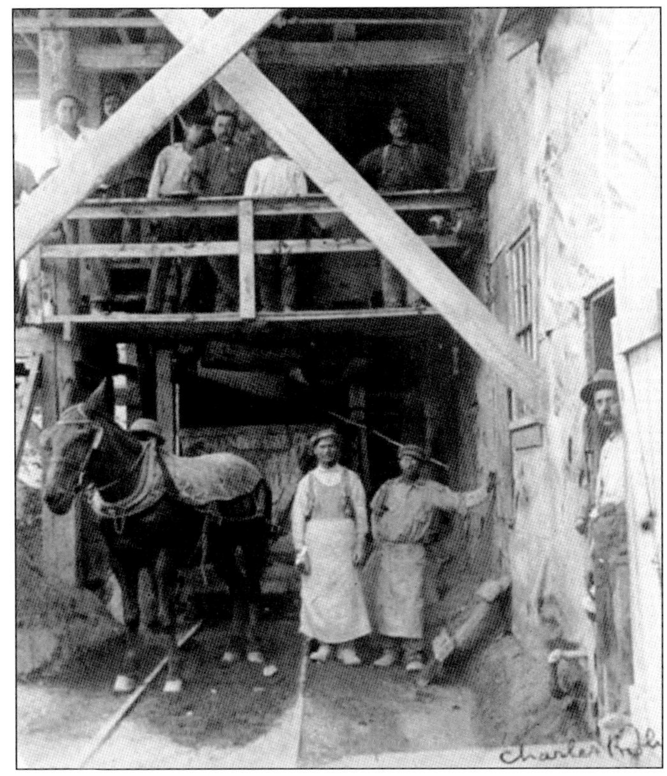

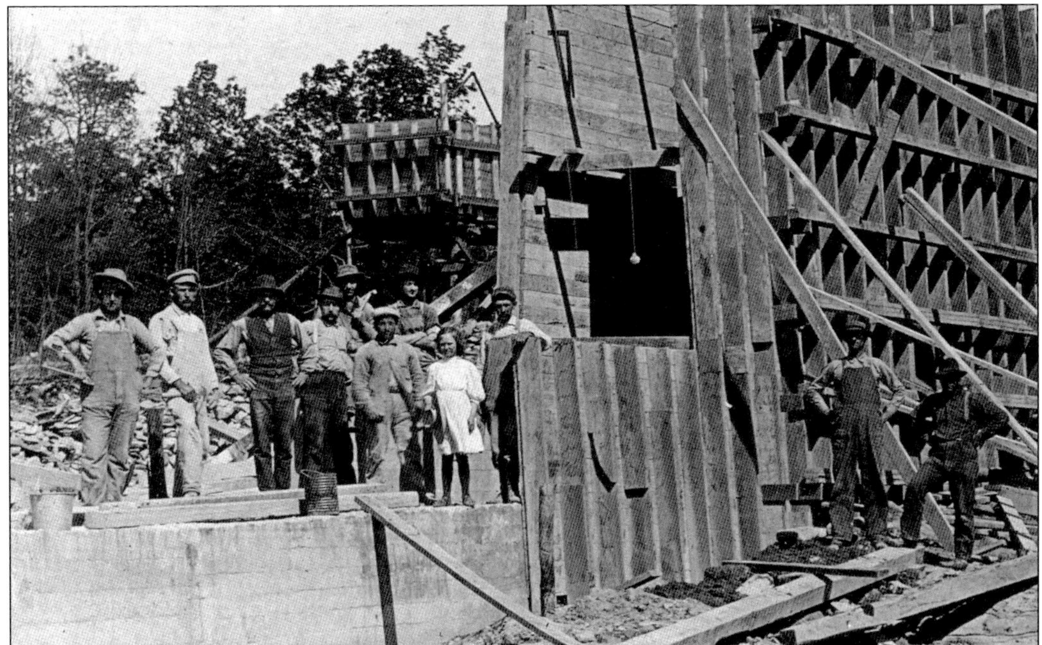

This crew is setting core wall forms. This wall is at least half height, with taller forms being added above the lower section. Forms gave the wall six or more feet of lift, typically 1,000 feet long. This wall continues to the left as the crew (with a helper) stands on a finished lower section. Concrete was delivered from a nearby plant by cableway, as there were no tracks or derrick around the area.

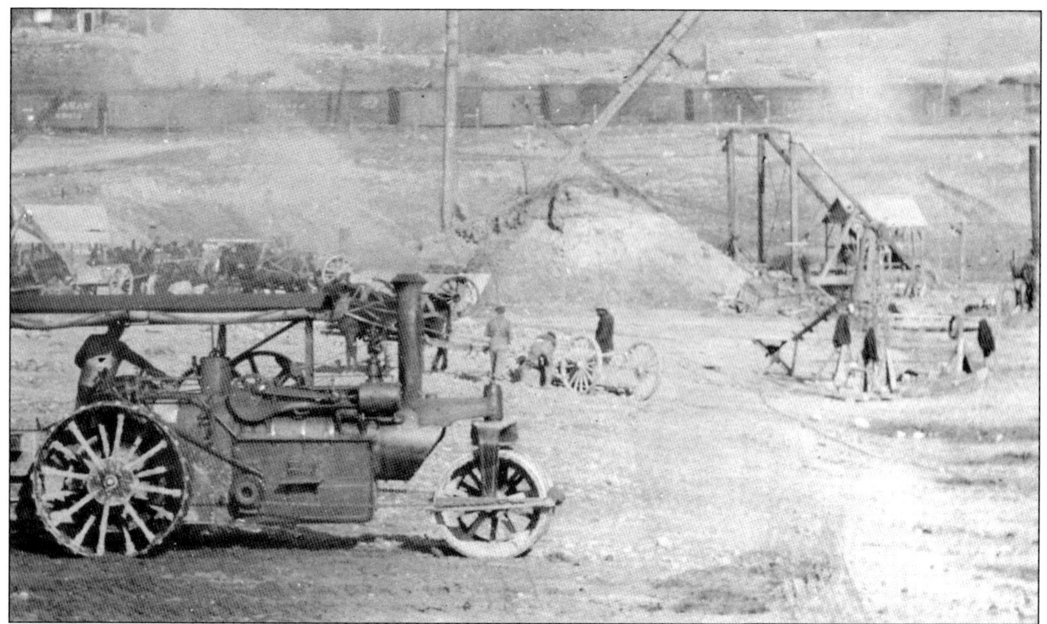

A worker steers a steam-driven Monarch and Kelly 12-ton roller to compact new earth on the embankment. The core wall is behind the roller. Coal and water are needed to keep this machine working along with water for the horse and men. The crew beyond the roller are adjusting a horse-drawn scraper to level newly dumped earth.

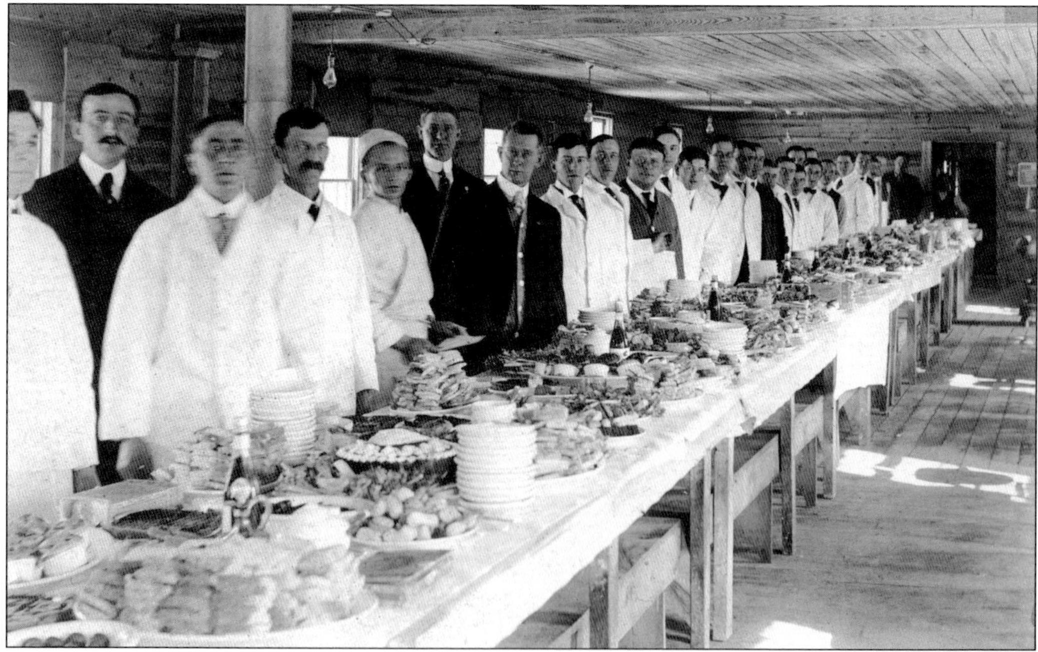

The staff and others prepare lunch for 500 engineers when the American Society of Civil Engineers visited the reservoir in December 1913 to view the construction methods and progress. The Friday or Saturday trip concluded a weeklong New York City conference. Travel was a round-trip steamboat ride to Kingston and then a round trip by a special U&D train, with lunch at the dining hall. (Courtesy of David Turner.)

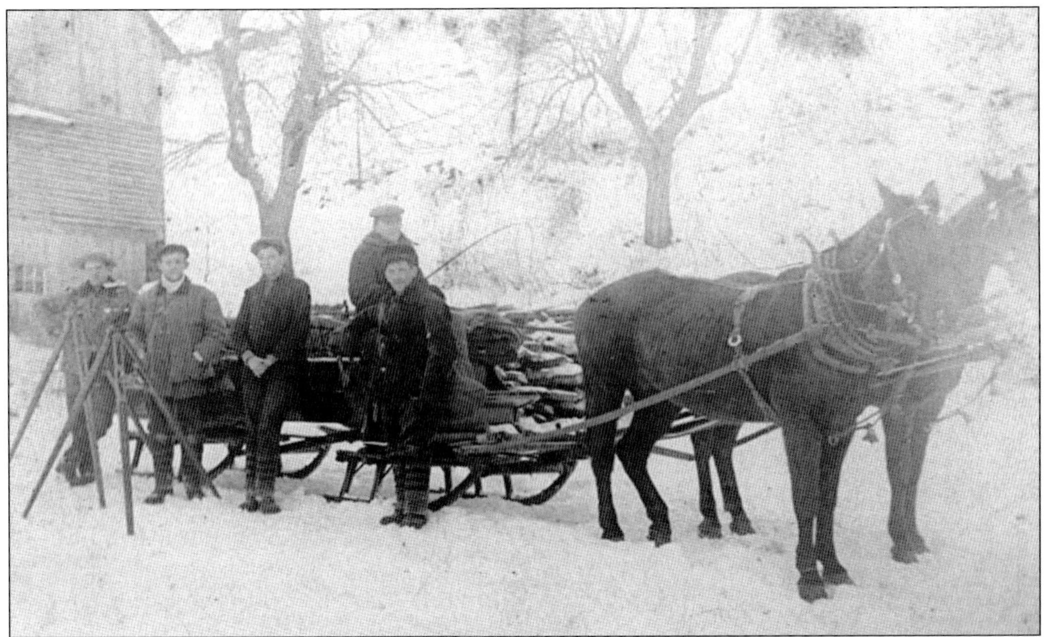

Surveyors were probably the hardiest of BWS employees. Their job was to know the land and where in relation to a tower on Winchell's Hill, near the center of the site, everything was located. Winters were severe, and outdoor clothing was not as good as what is available today. Precision was still important, even when feet and fingers were cold.

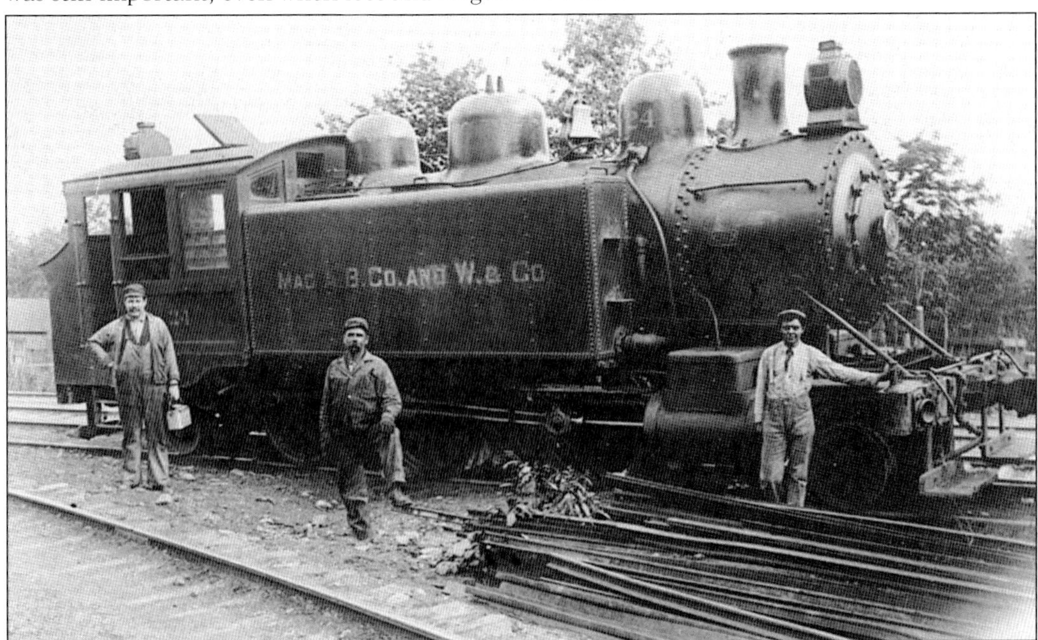

When the work of the crew of locomotive No. 24 was done at Kensico reservoir, they took it to the Ashokan site pulling a string of loaded boxcars. The route included a westbound trip over the Poughkeepsie railroad bridge and a few days of travel. Abandoned after a 1974 fire, the mile-long bridge opened for foot and bike traffic in 2009 as Walkway over the Hudson, a state park. (Courtesy of John Ham.)

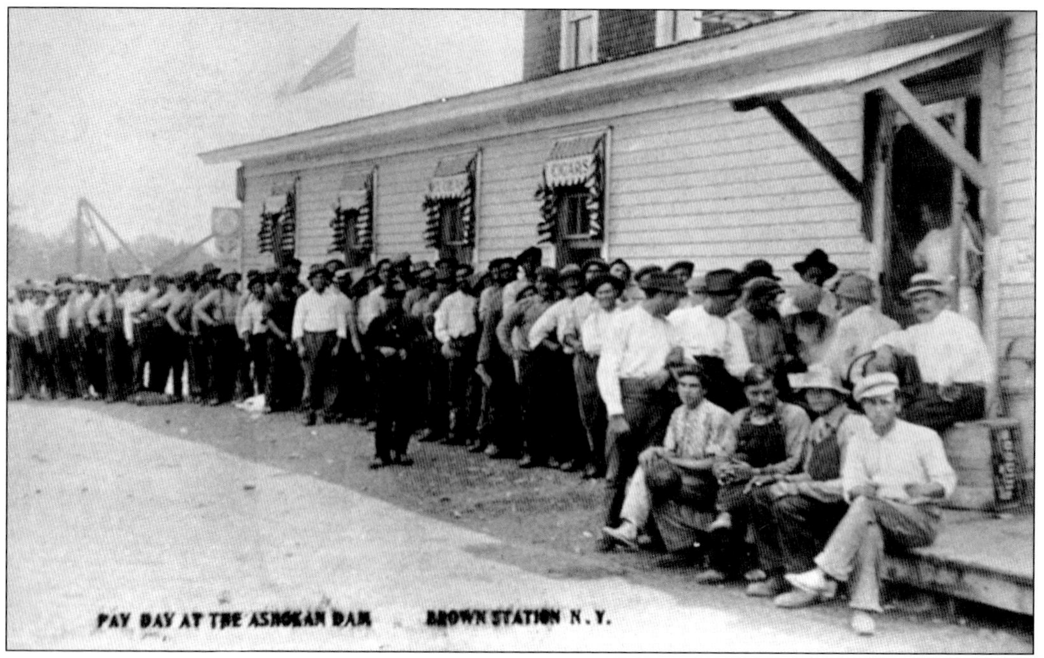

Most important to all workers is payday. The paymaster operated out of a wing of the Ashokan National Bank, seen in chapter seven. As stated in the booklet *Ashokan*, most workers deposited their checks into bank accounts and used Winston brass for day-to-day spending. Images of the one-inch, one- and ten-unit tokens are shown below. Their buying power is unknown.

Typical Wages

Most references state wages, all are similar numbers

Job	Wage	Period
Laborer	\$1.20 - \$1.60	daily
Plumbers	\$2.00 - \$2.40	"
Carpenters	\$2.50 - \$3.00	"
Stoneworkers	\$3.00	"
Powderman	\$10.50	weekly
Shovel operator	\$125	monthy
Helper	\$75 - \$90	"
Dinky (NG RR)	\$90	"
Patrolman	\$75	"

Many other project jobs - no descriptive data

Company brass tokens

This table with the Ashokan Bank in the background provides the only wage information in print. Most references only show a few jobs. White provides aqueduct workers' wages, above ground and tunneling, similar to those paid by Winston & Co. One job not listed but pictured in Diane Galusha's *Liquid Assets* was a water boy, who was paid 10¢ per hour. White's footnote says wages varied by year—did the water boy get a raise too?

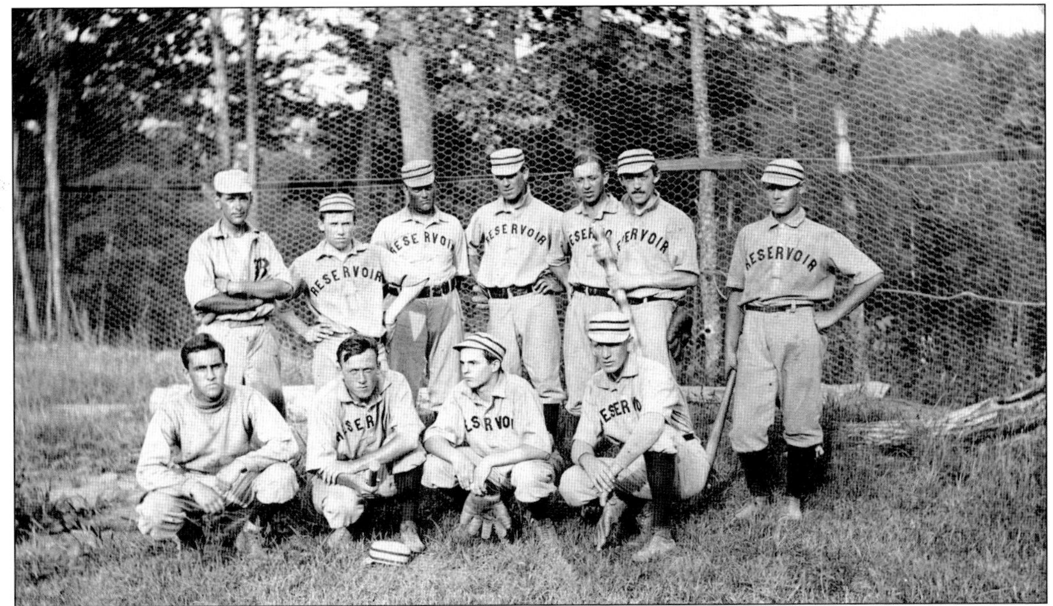

Time off was usually used to enjoy a ball game. The camp had a ball field that was used almost every evening. Younger skilled workers assembled a team to play against local towns' teams. In 1909, Ashokan played Olive Bridge and Shokan for a silver cup and won. The BWS engineers team won a silver cup over Ashokan. The Italian portion of the camp always has a number of bocce games in the evenings.

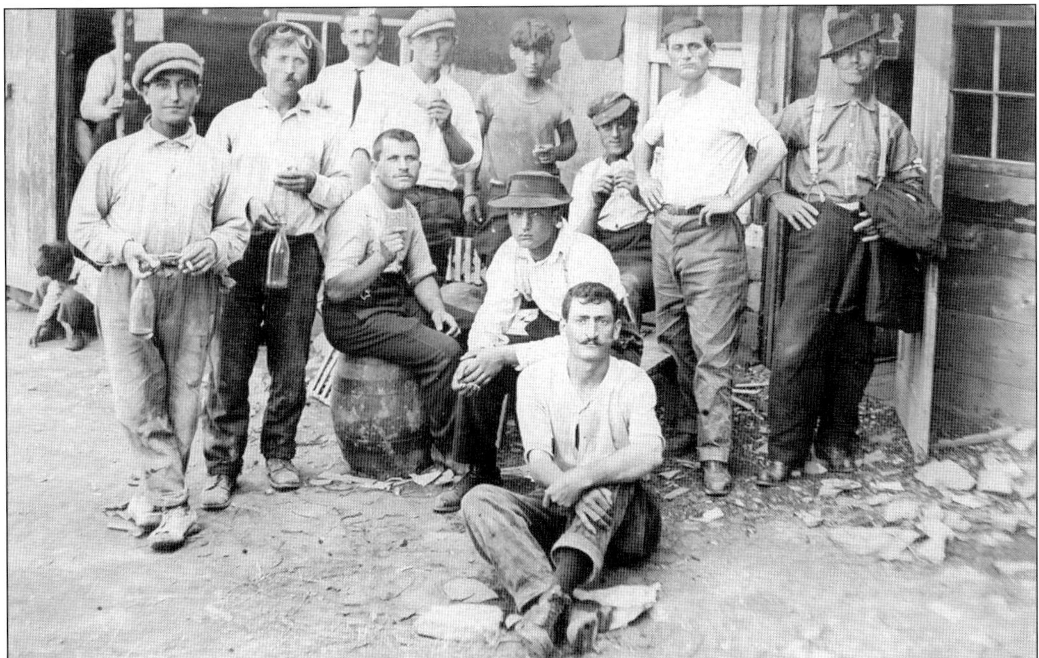

It looks like an after-work or weekend card game is being interrupted by the photographer. The man on the barrel holds bills while two others hold cards. Some writers mention that card games were played by the workers, although cards were not sold at the commissary. Beer was not sold on the site, so this photograph may have been taken off BWS property.

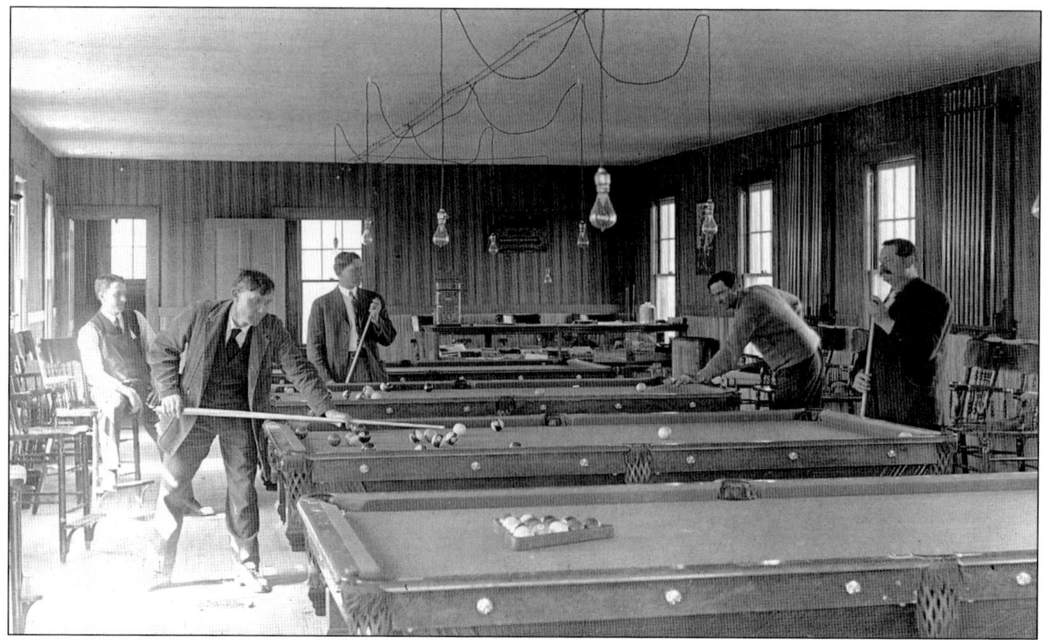

A favorite pastime for many was found upstairs at the bank building. A subcontractor installed a pool hall that did well and kept the workers entertained during off-hours. Note the "casual" dress code for men who worked hard eight hours each day. The whistle blew at 8:00 a.m. when the job started, at noon, and again at 5:00 p.m. (Courtesy of David Turner.)

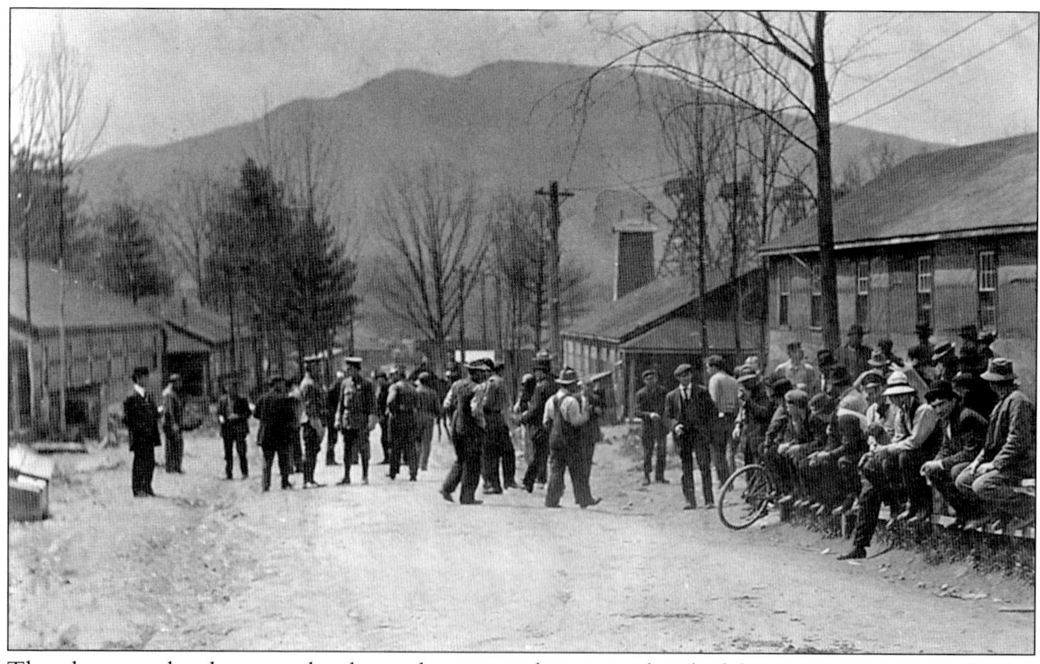

The photographer happened to be at the camp when some kind of disagreement occurred. Two BWS policemen are shown, along with others milling about. One of the four camp boardinghouses is on the right. The four cableways that stretched across the dam are visible at the end of the street. High Point peak is in the background.

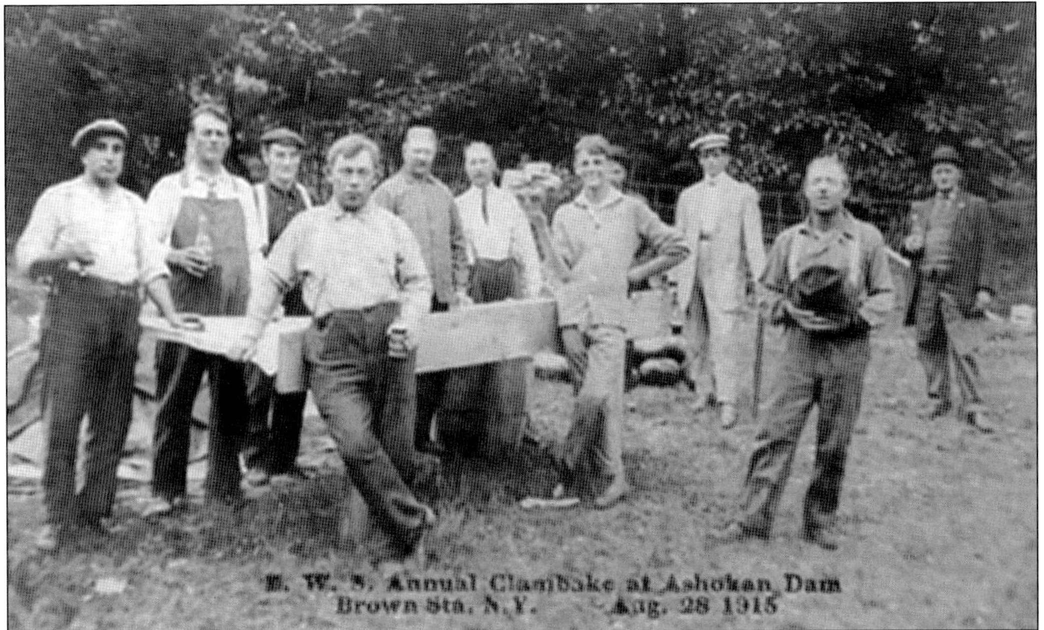

BWS treated its employees well. Here, they are wrapping up a clambake. The garb of two men at back right is interesting. They do not look like they are going to spend this Saturday afternoon with their co-workers. Their plan was probably to get the train to Kingston. The timetable on page 13 shows a cost of approximately $1 for a round-trip ticket to Kingston.

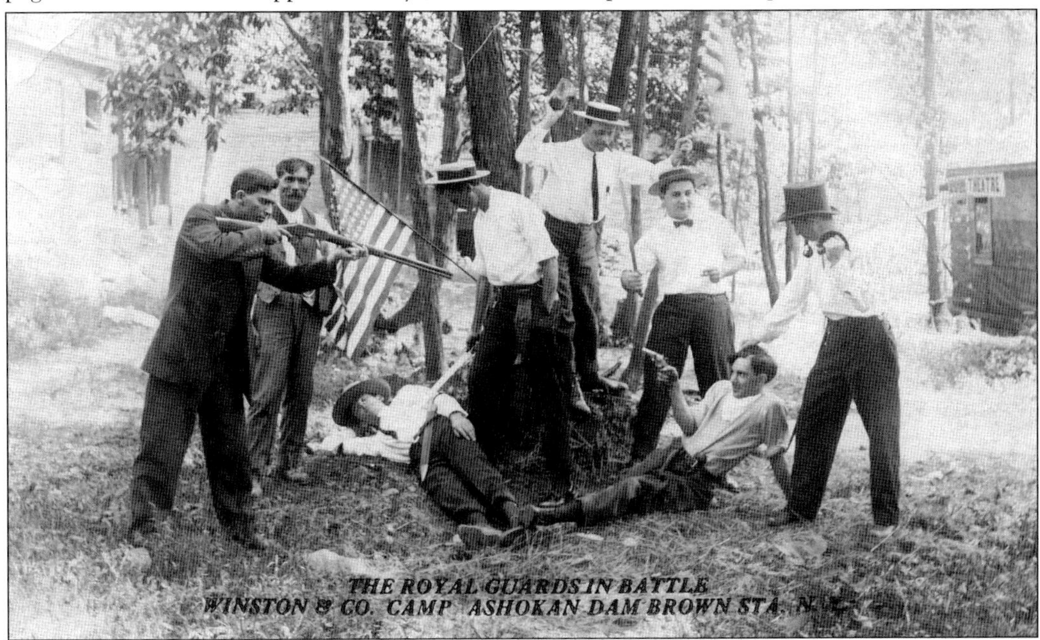

With over 8,000 employees in 1910, many with families, a little theater group formed. Note the box-office sign. The building was next to St. Peters Church and was used for entertainment and dances. Creating plays that put day-to-day activities aside made for good entertainment, as shown in another picture by Longyear. When construction ceased, the Longyear studio in Kingston was moved far away from the U&D depot. (Courtesy of David Turner.)

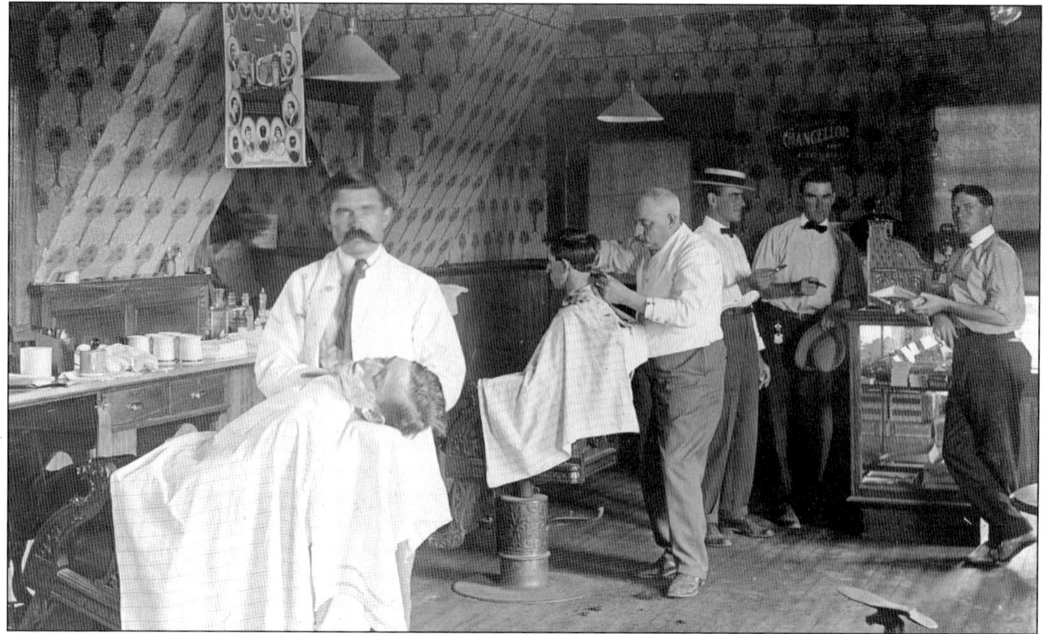

Workers needed haircuts, even while building the reservoir. Barbers did dress better in the early 1900s. There is no indication of what a haircut cost, but with wages in the $1–2 per day range for laborers, it was probably less than a quarter. The shop was on the bank's second floor, along with steaming baths. This was a busy shop, with the cigar stand. (Courtesy of David Turner.)

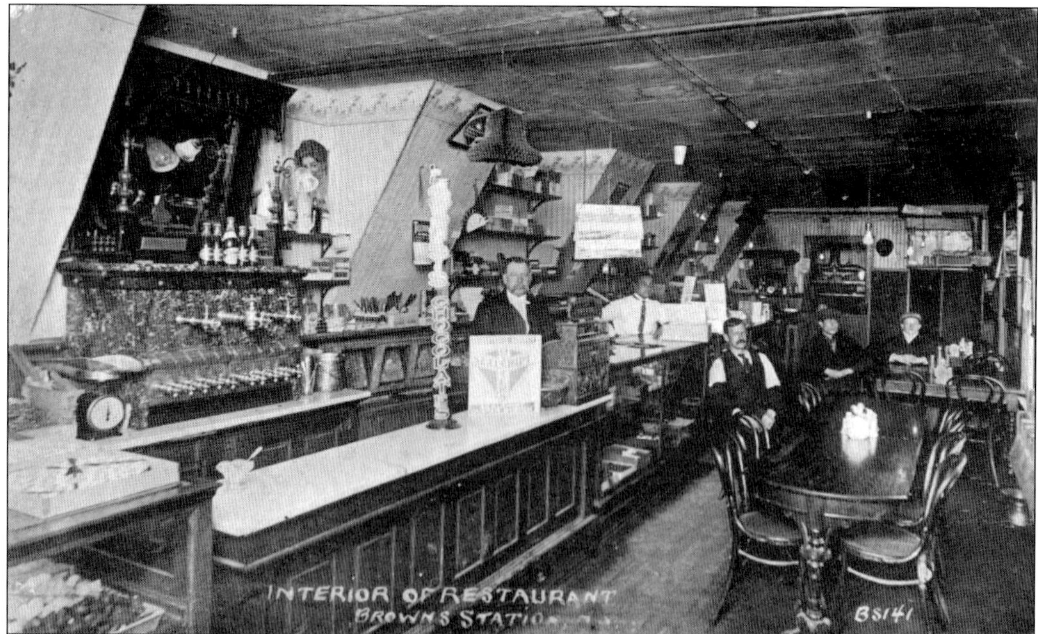

Also upstairs at the bank was a small restaurant. This is part of a series of 50 reservoir postcards made early in construction. The tall sign on the counter reads "Chocolates," and the one beside it is for "SealShipt Oysters," with a cross through the "S." This image must have been captured near Christmas, as a foldout paper bell is hanging above the counterman. Bare lightbulbs enhance nighttime dining.

Six

Workers' Camps and Public Buildings

Winston & Co. provided living quarters for all employees in a new village near Brown Station. There was a variety of housing for families and single men. Base dorm-type room and board was $20 per month. Rates for other accommodations are unknown. A camp bungalow could be purchased for $400. All necessary family food was obtained at the commissary. The contractors' railroad included a double set of tracks that ran through the upper part of the village to service the power plant and cement plant as well as supplies for the residents' well-being. The streets, all dirt, had lighting, as did all buildings. A kitchen with running water and full or half baths was provided in all accommodations. The village had a complete modern sewage system. All housing was weather-tight and had wood-burning stoves for winter heating. Rent included water, wood, and household trash pickup. There were dormitory buildings for single men with two or more beds to a room. The village complex also had fire hydrants supplied with water from a large reservoir at the top of Winchell Hill. Also on Winchell Hill was the bank, with the pool hall upstairs and a small restaurant. The commissary (or general store) had everything the workers needed, including work clothes and boots. A short way up the street was the hospital and nurses' quarters with a doctor, assistant, and three nurses.

Services provided for the workers were many. There was a barbershop in the bank building, a shoemaker in the commissary, a kitchen and dining hall for three daily meals, two school buildings, three churches, a police station, a bakery, and the contractors' office building. BWS had its own office building, bungalow lodging for employees, and an engineers' club in an undefined location away from the contractor's camp.

Some historians have stated BWS rented buildings "sold" to New York City. When water levels were nearing full pool on December 12, 1916, everything was out of the basin. Workers' accommodations were demolished. There was little need for BWS employees.

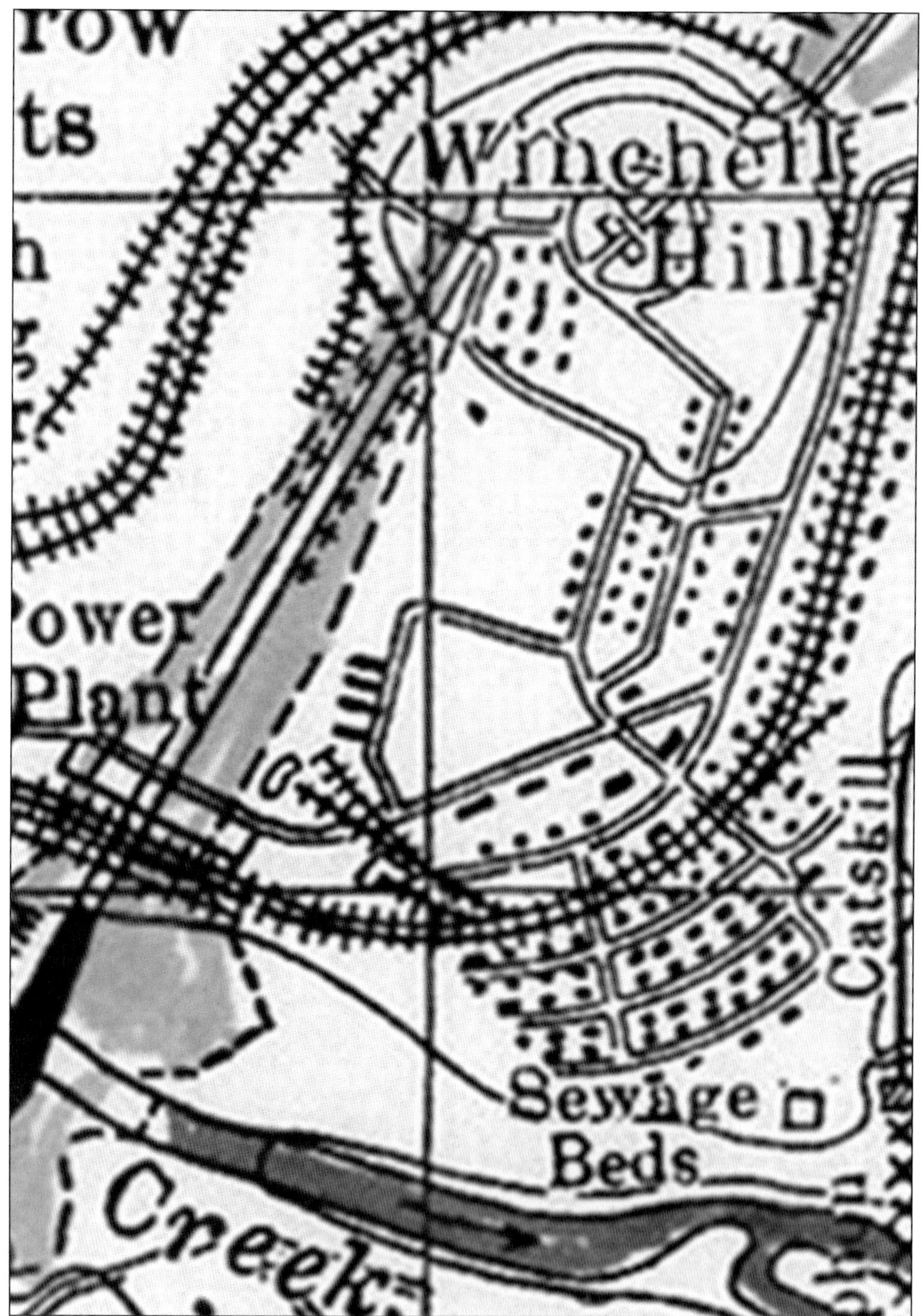

This camp map is a detail from White's construction site map. There is no indication of what style cottage was built where, or the location of public buildings. These were probably near the east-west road adjacent to the double-line railroad tracks. The railroad was needed to supply all those buildings frequented by workers and families, including the service buildings, cookhouse, and bakery.

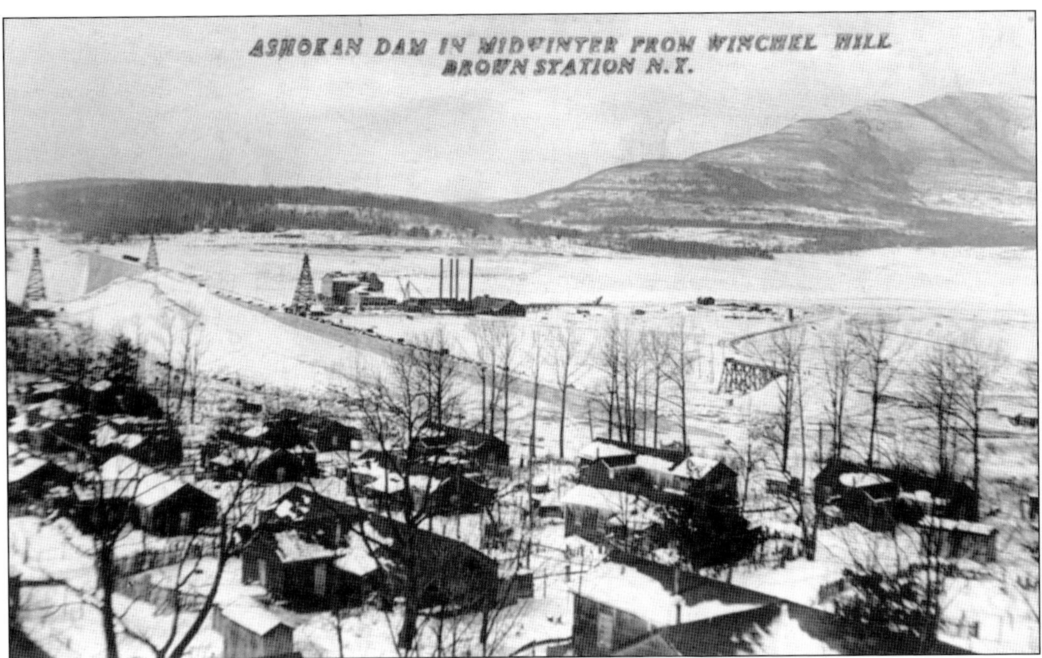

This wintertime image shows the western half of the camp from Winchell Hill. The power plant and stone crusher are in the distance. A variety of residences can be seen on the hillside. The wide grey line from the dam is the North Wing core wall. The embankments would be increased when work commenced after the spring thaw.

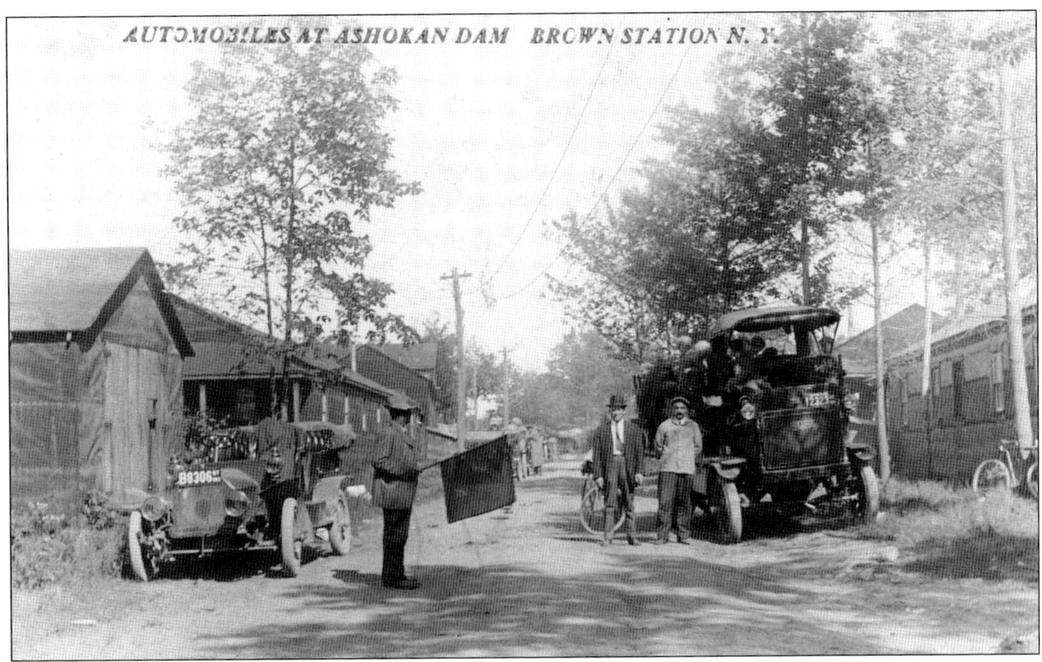

A flagman stops vehicles while a work train passes through the camp. The contractor's railroad had a double track through the upper part of the village, where trains ran at any time. The image is from 1910 or later. The larger vehicle looks like a bus.

Camp Buildings

1 special 3 room dwelling with bath
7 special 4 room dwellings with bath
7 special 5 room dwellings with bath
7 special 6 room dwellings with bath
1 special 7 room dwelling with bath
21 standard 3 room dwellings, of which 4 have baths
21 standard 4 room dwellings, of which 5 have baths
1 standard 3 room dwelling for foreigners
81 standard 4 room dwellings for foreigners
11 standard 5 room dwellings for foreigners
4 standard 6 room dwellings for foreigners
20 standard 3 room dwellings for colored people
4 standard 4 room dwellings for colored people
4 barrack buildings, for men only, of 4 rooms each
4 barrack buildings, for men only, of 9 rooms each

MASSACHUSETTS AVE BROWN STATION N.Y.

This inventory list of dwellings and construction detail is taken from the *Ashokan* booklet. The building inventory is highly varied. All were built on concrete foundations of two-by-four and two-by-six construction with front and rear porches. One-inch boards sheathed the outside. Interiors were tongue and groove pine or hemlock walls and double layer floors with paper wind stop. A paper-like covering wrapped the buildings.

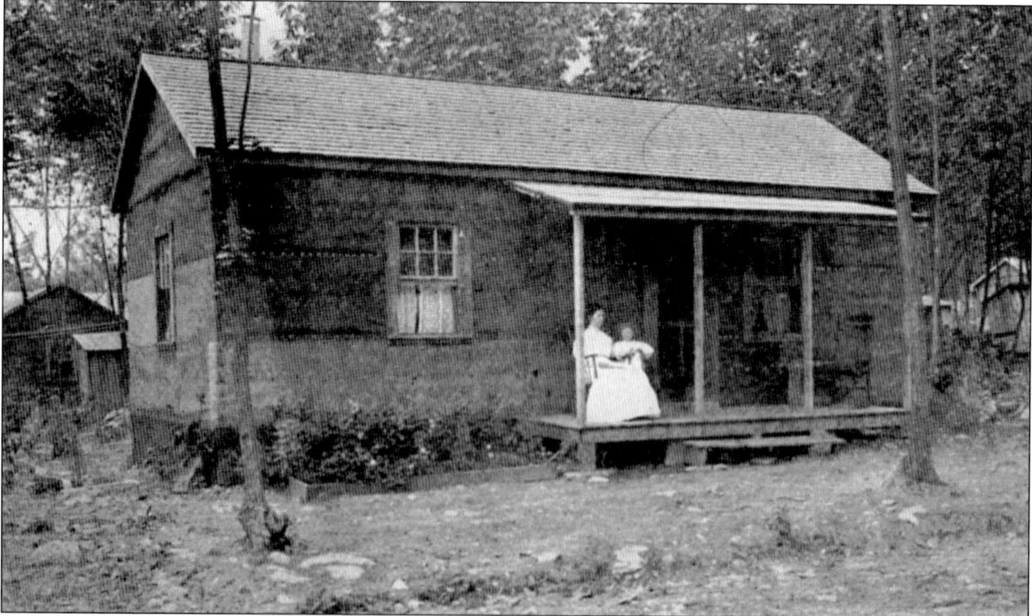

The standard three-room dwelling had a sitting room, bedroom, and kitchen with running water. The privy was outside and connected to the sewer system, as can be seen at left. The kitchen water drained through the privy. All buildings, private and public, were built on concrete foundations with a wrap of weather-resistant material. (Courtesy of NYPL digital library.)

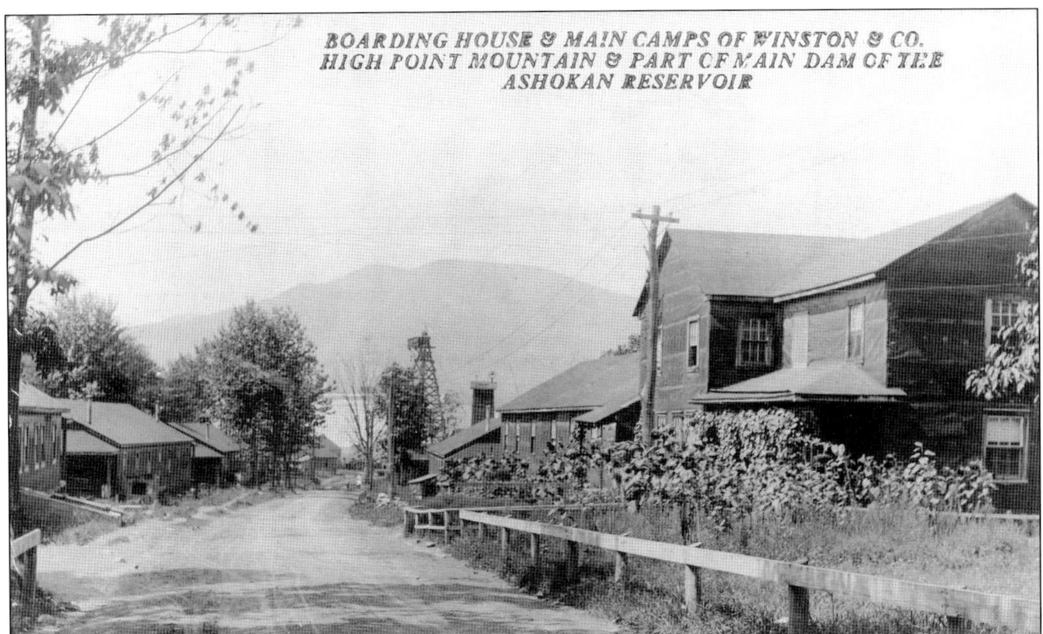

This two-story boardinghouse was on one of the east-west downhill roads that pointed toward the main dam. The dwelling was for single men, with mostly two to a room, and one of many throughout the village. Their monthly rent paid for a cook and housekeeping. The white mass in the background is the dam. Only one of the cableway towers remains.

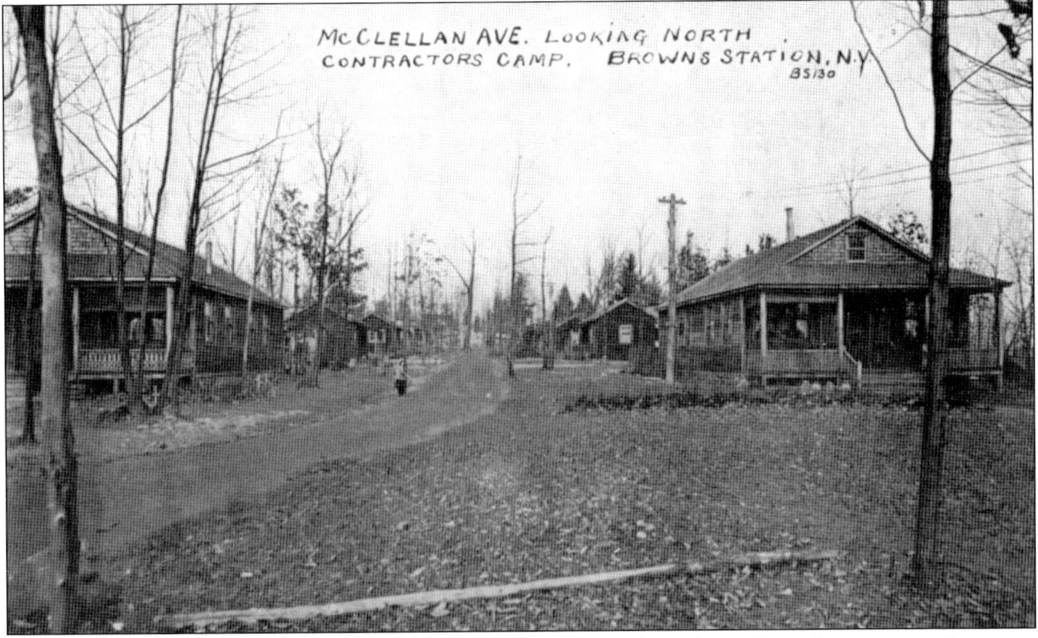

McClellan Avenue is one of the few identified roads within the village but not on the map. Dwellings on either side of the road appear to be similar, probably housing a larger working family or a group of men. Farther up the hill would be the tracks of the contractor's railroad. The roofs of all the buildings were covered with weather-resistant rolled material.

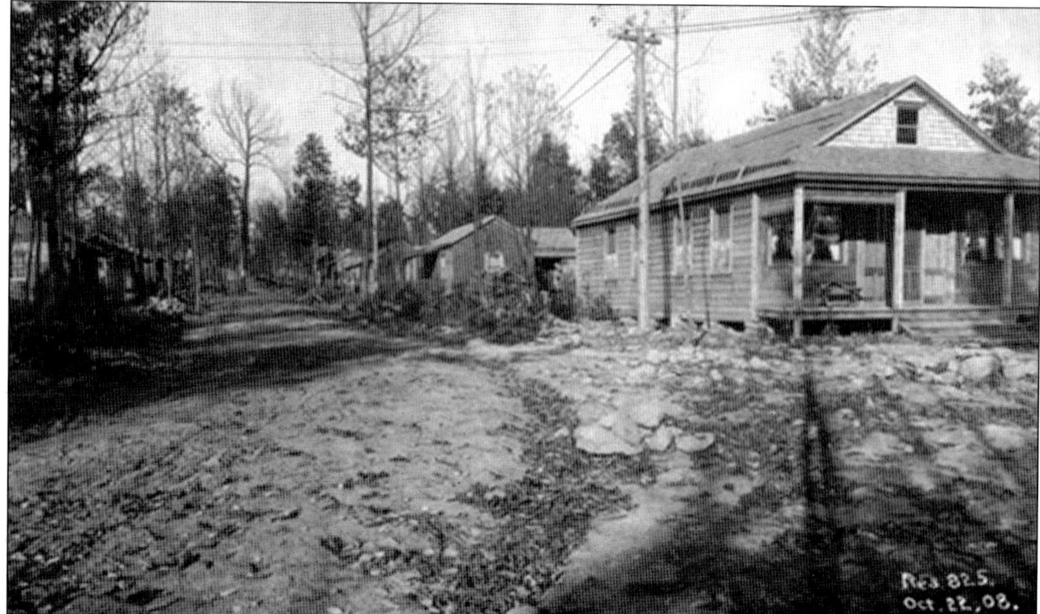

This photograph appears to show the same place as the previous image, used here to show the date it was made. Considering Winston's contract started in September 1907, this photograph shows a complete camp, including electric power to most dwellings in just more than one year. That work included clearing land, installing water and sewer lines below three feet, making the foundations, and building each dwelling. (Courtesy of NYPL digital library.)

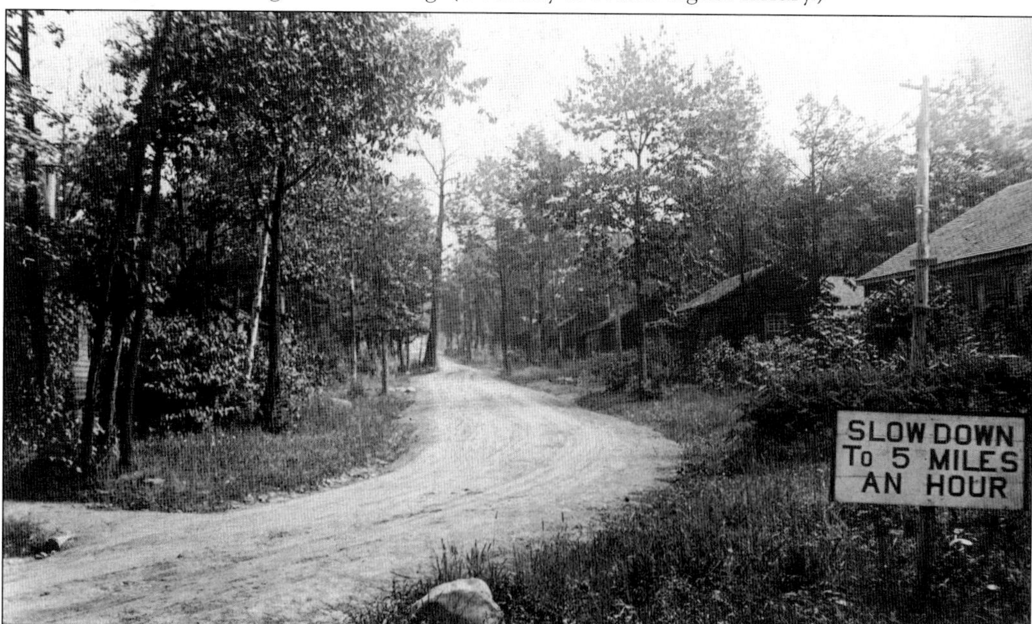

In later years, McClellan Avenue must have become a major commuter road, as a speed-limit sign was installed. The tenants of these cottages were supplied with daily trash service, and firewood and ash pickup in the winter. Winston & Co. supplied two wagons and workers for this assignment. McClellan Avenue must have been easy for the photographer to reach, as many postcards feature this corner.

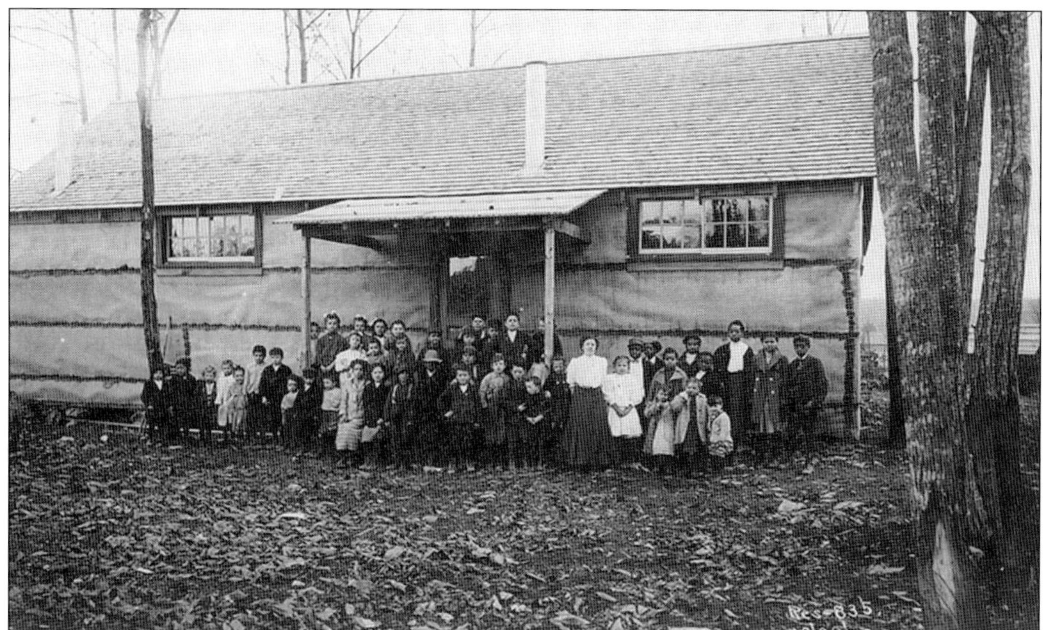

This school for workers' children had to be erected due to overcrowding in the Brown Station building. This school on the campgrounds opened in the fall of 1909 with an additional teacher hired. In the evenings, the Italian Immigrants Society used the school to teach English to all interested workers. The society held Kindergarten for children three and a half to six years old in the other half of the building.

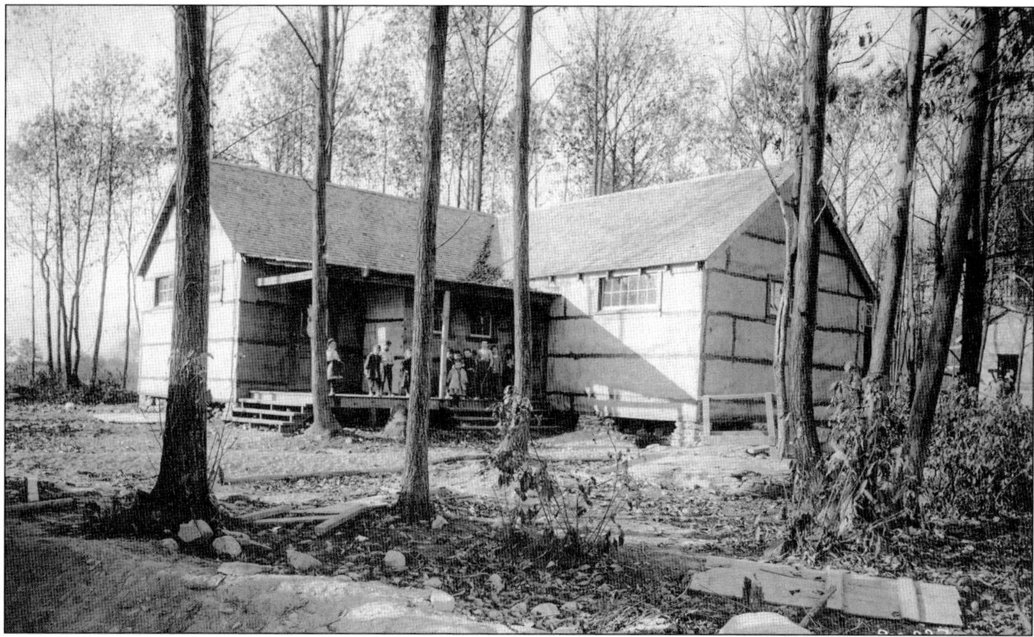

A second school was built in 1911 in an area of the camp selected to shorten the walk to school. Buildings were erected in the same manner as the homes and heated with wood. In the winter, snow was banked against the foundations to keep the floor warmer. Children of workers in the dike camps attended schools in the nearby communities. (Courtesy of NYPL digital library.)

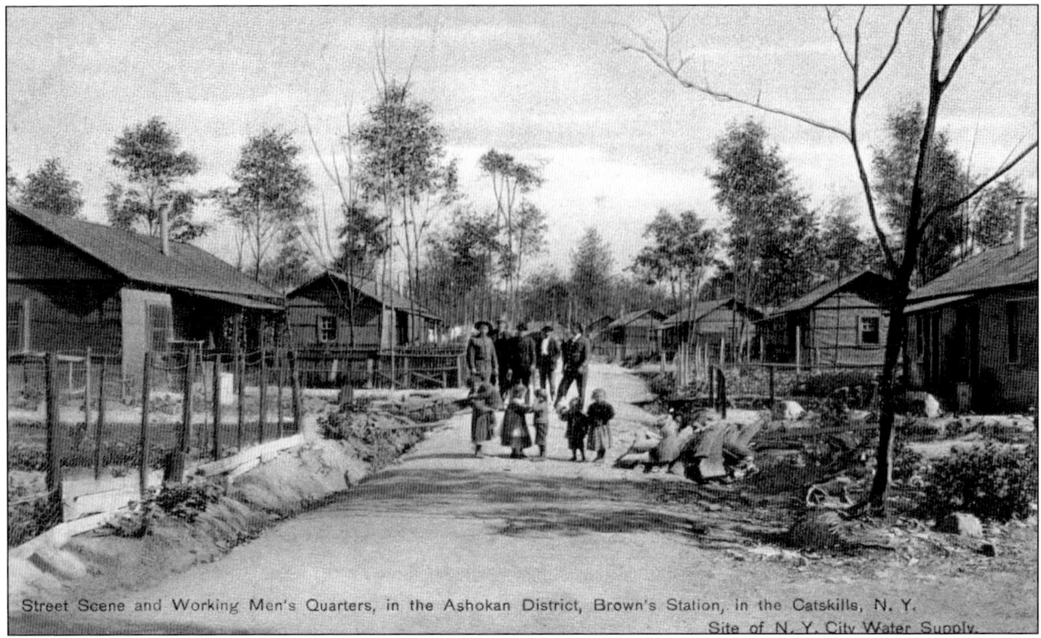

This popular postcard shows children with adults behind them and a local BWS patrolman. Some cottages along the road have installed fencing, primarily for their gardens. It looks like the area had a heavy rainstorm, as a new ditch has been made on the left of the road, and what looks like sandbags are on the right.

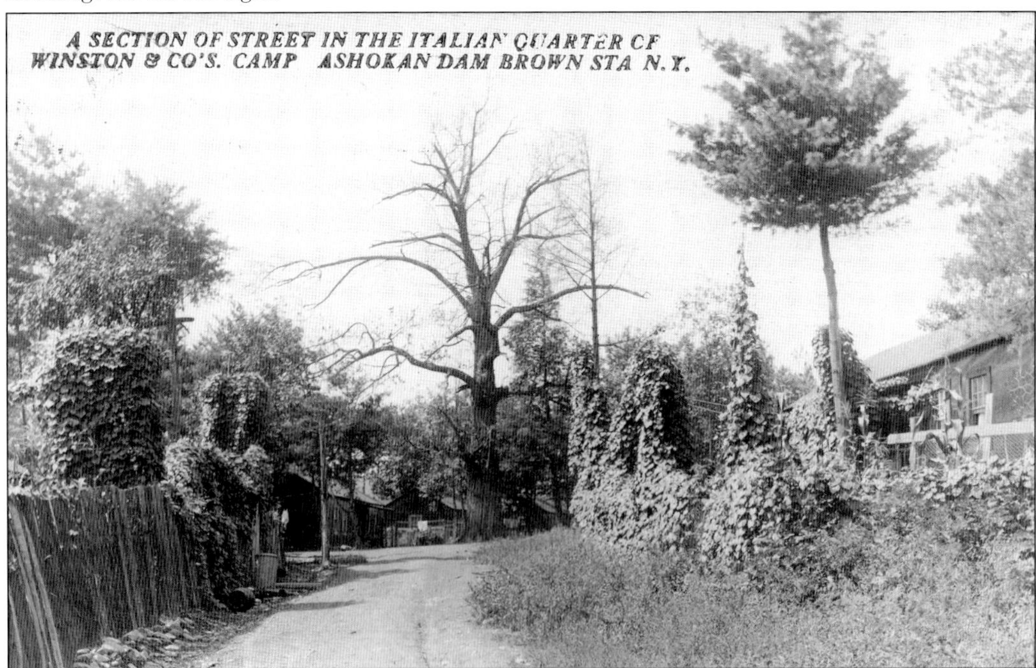

There is no indication where the Italian section was located; however, if one could walk the camp, the number of plantings would be an indicator. The photograph appears to show pole beans or other vining crops in a home garden. The Esopus basin was a glacial lake with many feet of sediment under the soil surface. Pre-reservoir residents farmed this soil. Many workers had gardens.

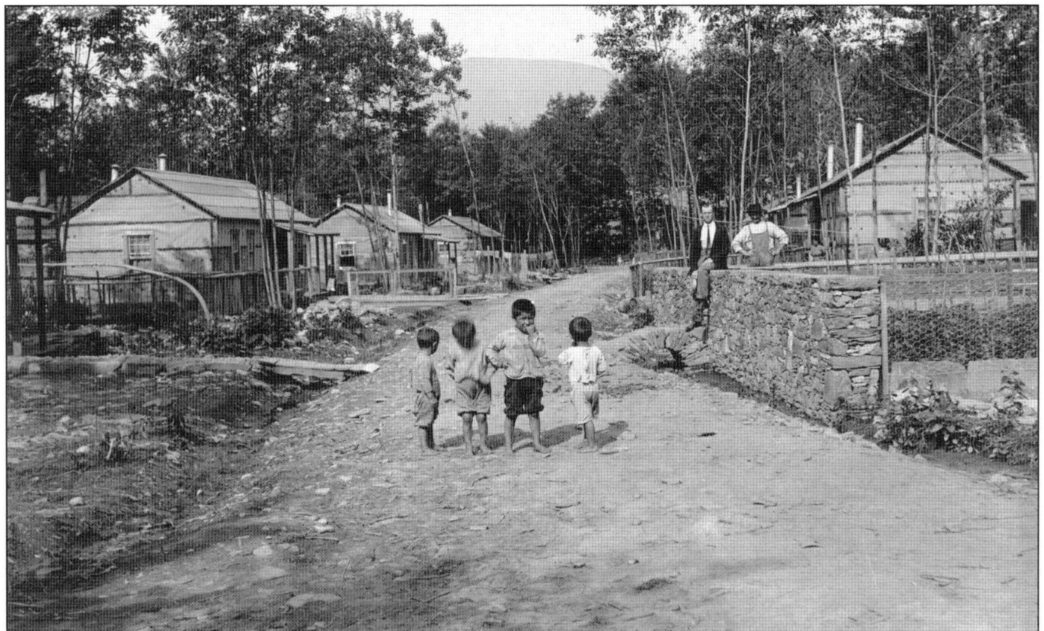

This camp postcard appears to show the same road as the upper image on the facing page. The children are dressed for a summer day, and the cottage on the right now has a dry-laid stone wall. The man in the jacket and tie is with a worker who is probably working a late shift.

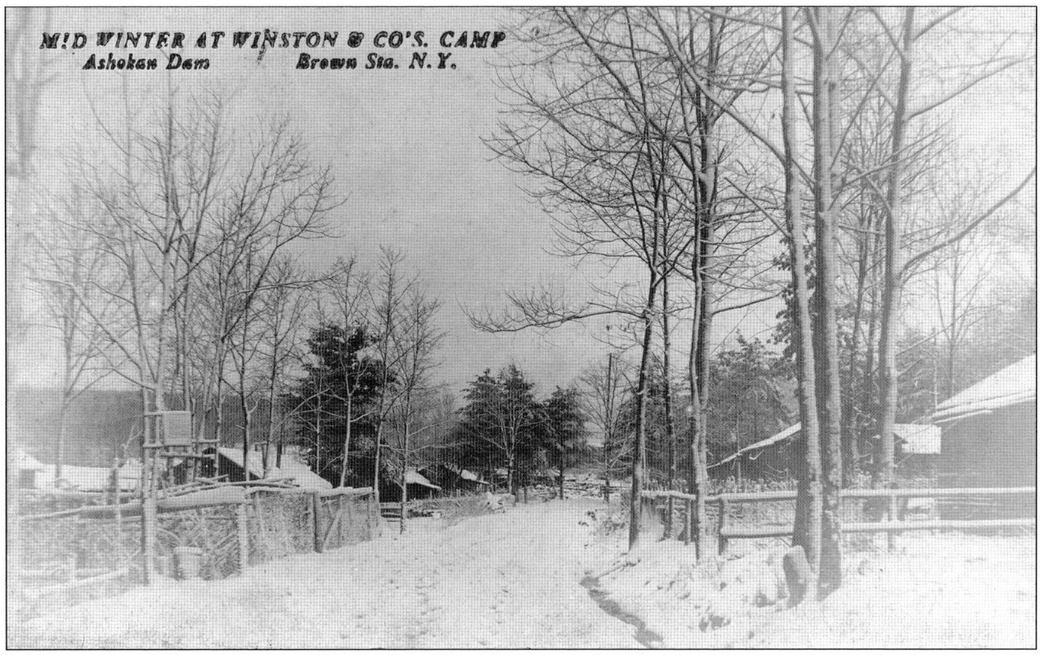

This postcard shows a typical camp road on a winter day. The snow appears to be fresh, as the roofs are covered. Most workers spent the winters at camp unless they had relatives nearby. Note the many footprints in the snow. This is another Longyear card, so he did get to the camp on winter days.

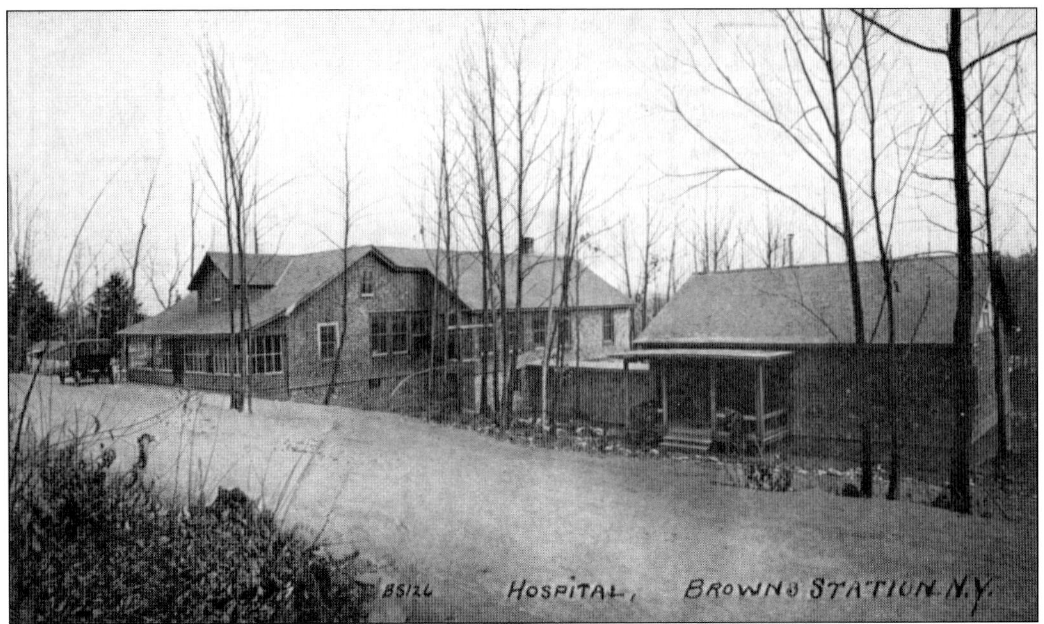

This postcard shows the hospital beyond the nurses' and doctors' quarters. A doctor and assistant with three nurses were the medical staff. They were responsible for vaccinating all workers to meet state health standards as well as treat typical health problems. They provided emergency care for any injury. There had to be many, although none have been identified. The county coroner was contacted for any death.

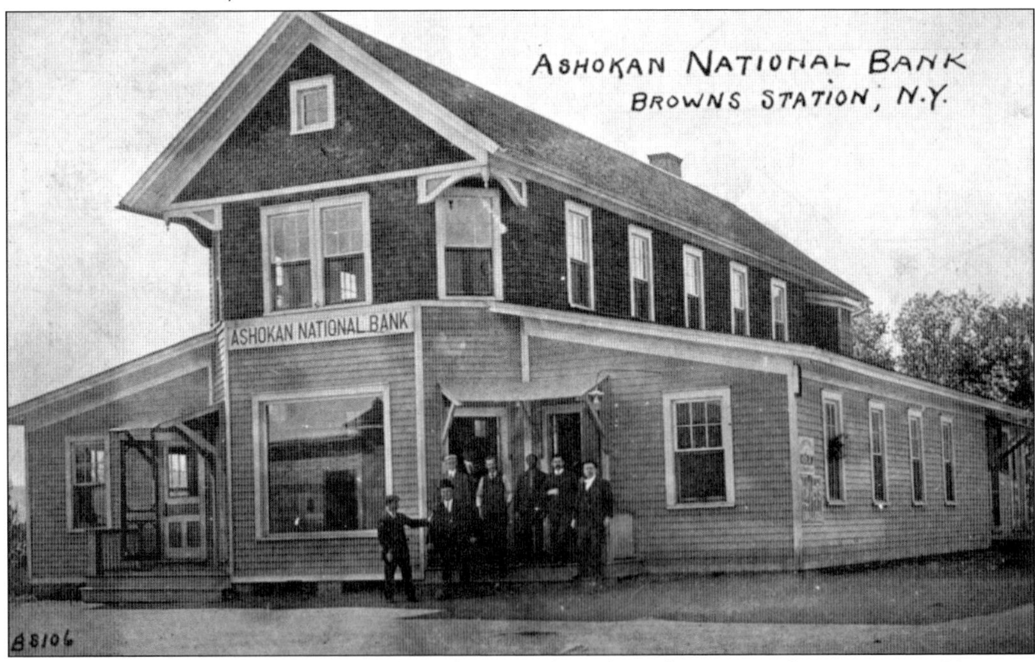

This bank image with staff appears to have been made early in the course of construction. Upstairs, the bank housed a barbershop, restaurant, and pool hall. The building was built by Winston & Co. as part of camp services and located near the double-rail line on the uphill side of the camp. A camp post office building was nearby to the left.

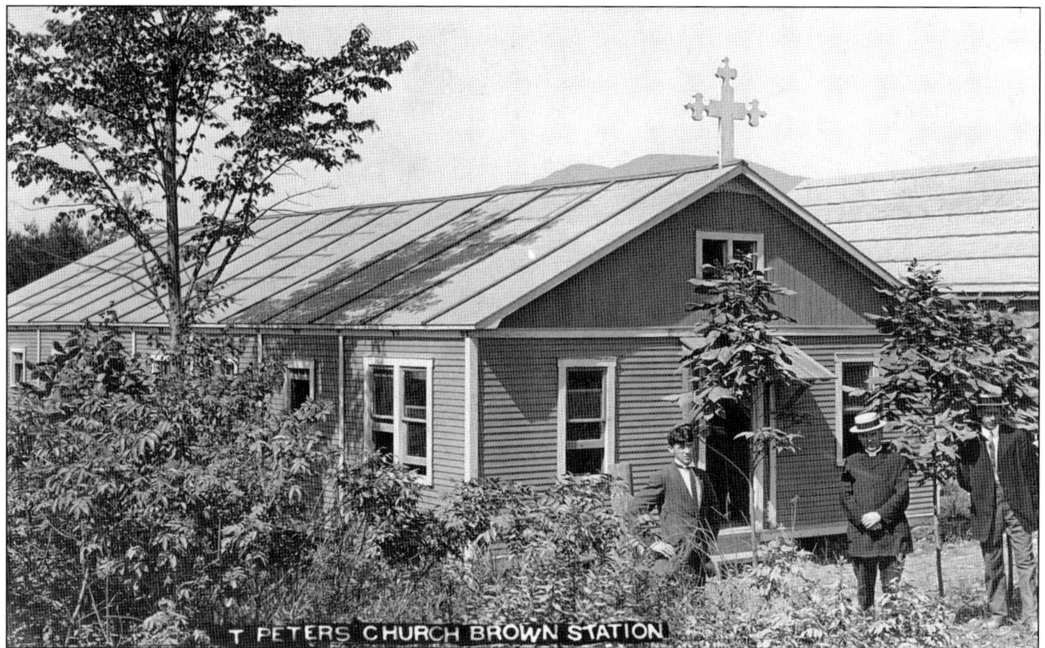

St. Peters Church and hall was near the top of the camp hill and was built by Winston & Co. The priest and two assistants are shown here. St. Peters was the largest church, as almost half the workers were Italian. The church hall (at right) had a double floor for dances and other entertainment. There was also a union and a black church at undefined locations.

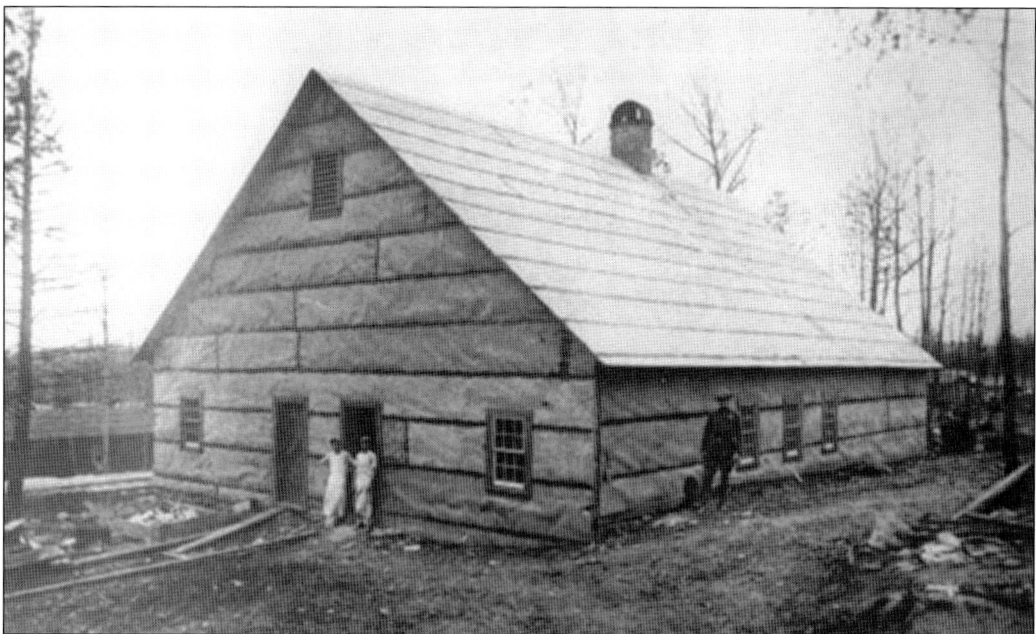

The cookhouse provided all the meals for the camp workers. The workers' dining hall was thought to be in the lower part of the camp near the barracks, like the single men's buildings. The bottom image on page 84 shows the dining-room staff setting up the dining hall for an engineers' visit. (Courtesy of NYPL digital library.)

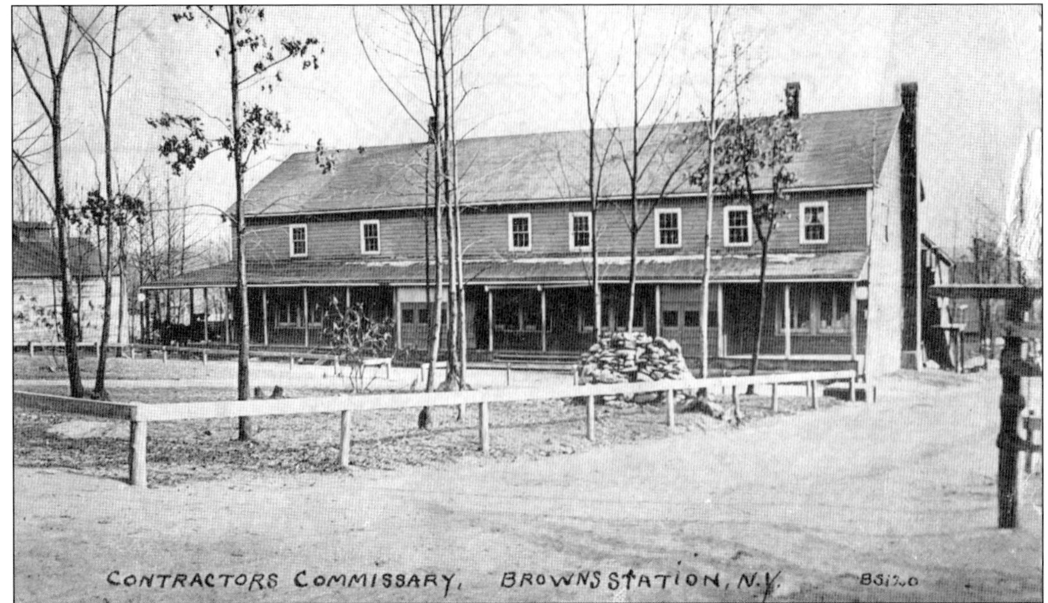

The commissary building, or general store, sold everything anyone at the site would need, except alcohol and cards. Vendors were not allowed on-site, but on weekends, they were outside on local roads. There was a large cellar under the commissary for cold storage, including an icehouse. According to the *Ashokan* booklet, during the summers, "the ice wagon makes two daily trips around the entire works." Ice cut in the winter was stored in off-site icehouses.

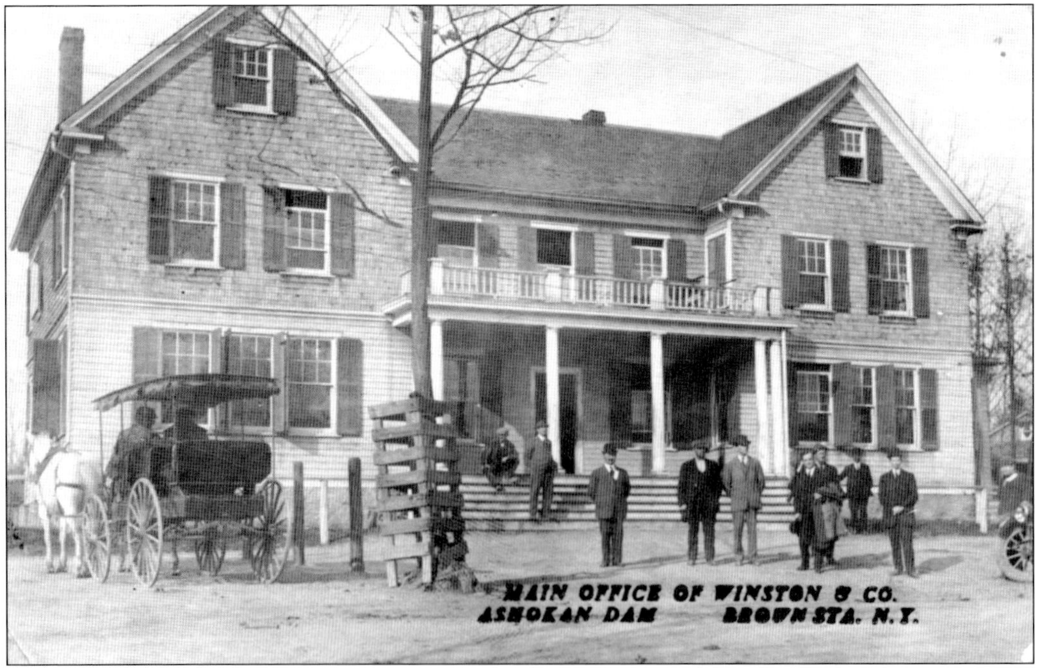

Winston & Co.'s office was a larger building than the bank, but it too was demolished when the project ended. The building contained all the construction documents, drawings, and specifications for the work. All of the firm's inspectors and section leaders had offices here. The people pictured were probably all key players, including James Winston, believed to be the man in the light suit.

Seven

Large and Small Equipment

The tools used by the workers varied from a pick and shovel and mule wagons to the Lidgerwood overhead cableway system run by skilled operators. The images on the next few pages show some of these in use or posed, depending on the speed of activity. Some tools, such as derricks with hoisting engines and narrow-gauge railways, were used from the beginning of the project even before the reservoir contracts were granted.

Portions of the coal-fired power plant came from the Kensico Dam early in the work. More boilers were added. Steam power was used for compressed air, which was used for power wherever it could be piped and for electrical generation. All the hoisting engines used with derricks, as well as the cableway motors, pumps, and rock drills, were powered by compressed air. In some locations, power shovels were driven with compressed air rather than steam. Though not mentioned directly, fixed facilities at the site seemed to be owned and operated by BWS. Stone crushers and the cement plant may have been BWS-supplied structures, as was the operation of the Yale Quarry. The dam construction contract does not mention coal costs, yet the contract paid $1.65 million for 1.1 million barrels of Portland cement.

Winston & Co. had nine American Locomotive Company (ALCO) saddle tank standard and 14 H.K. Porter narrow-gauge locomotives and skilled train drivers for most of them. There were 72 six-yard side-dump standard-gauge cars, primarily for moving embankment materials. To ride the top of core walls, Winston & Co. had 14 three-foot-gauge locomotives and 250 four-yard side-dump cars. It is still unknown if the firm sent its trains for cement or if it came by way of the U&D. Hauling earth for embankments was most of the work for the trains.

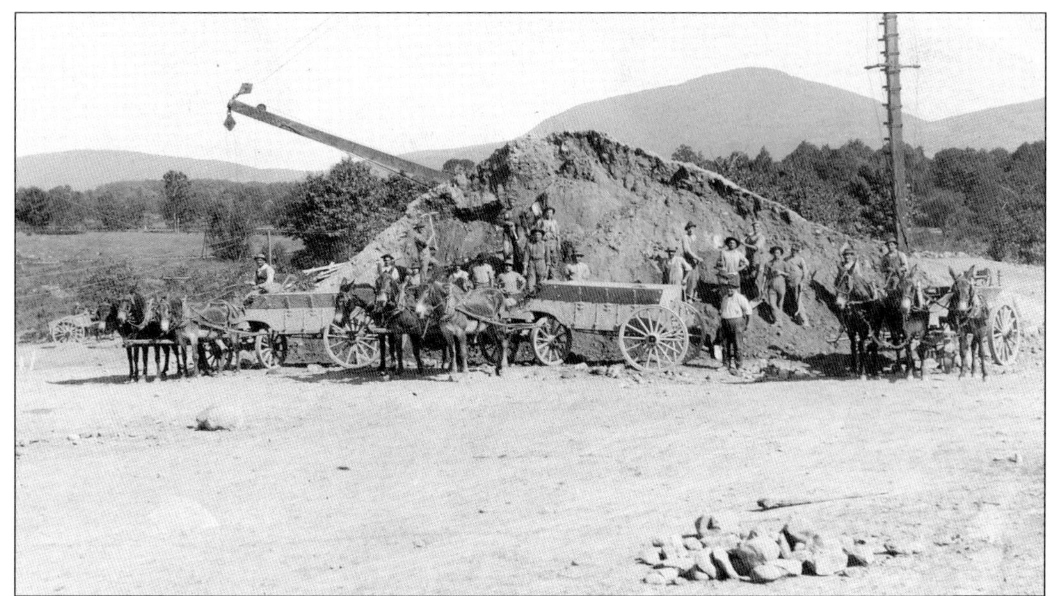

Mule teams were the trucker for moving anything, but mostly earth, primarily excavation and building embankments. Mules knew when the workday was over. According to C.S. Bergner, a child of a Winston supervisor, as reported in *York State Traditions* magazine, "Exactly at ten minutes of 5 . . . a weary mule . . . began to bray . . . all 365 mules were braying in unison. Their colored mule drivers standing up," racing to the barn.

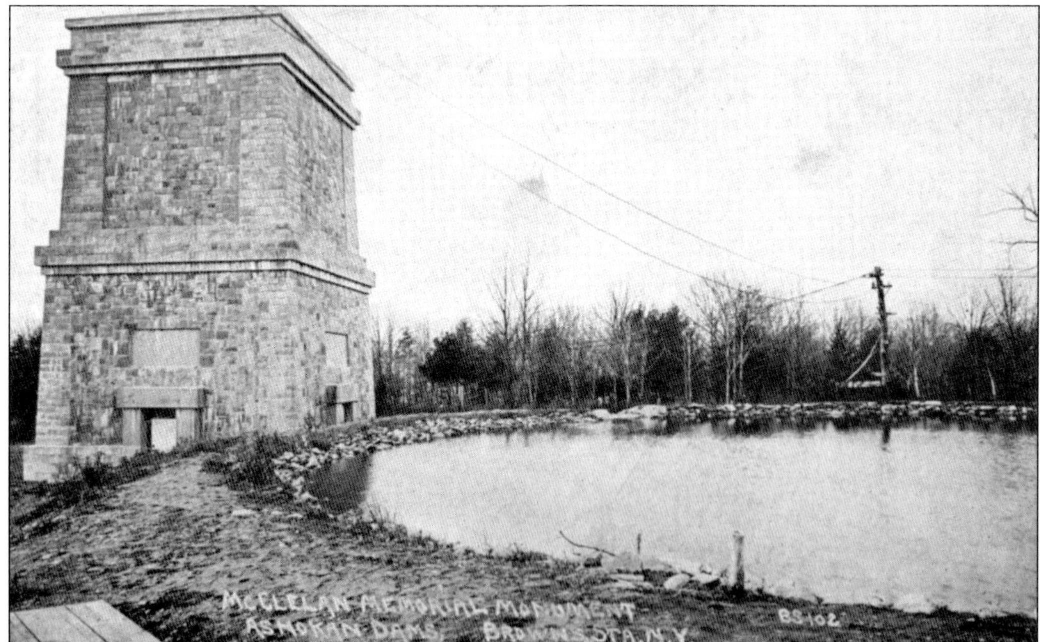

On the crest of Winchell Hill, BWS built McClellen's Tower, point zero for all survey work. The pond shown was the water supply for fire hydrants; it was gravity fed to the camp and all other populated areas of the site. There were two major fires: lightning struck the coal storage shed, but only the wood burned. The more severe one took place in the fall of 1914. The mule barn burned with all the animals inside; most were rescued, but many perished.

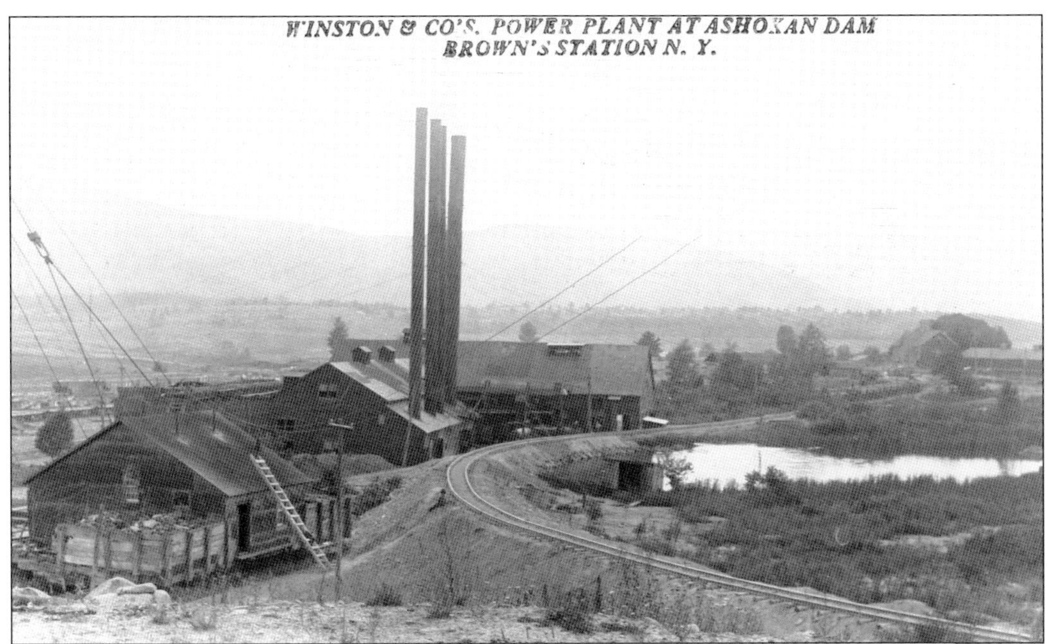

The power plant produced steam to make all the electricity and compressed air used on site. Five steam boilers drove the compressor plant and two generators for all site lighting. Significant overnight work was done to speed up construction. Cast blocks were moved from the block yard, and large boulders were placed where they would be needed by derricks, all under lights. The plant burned 1,400 tons of coal per month.

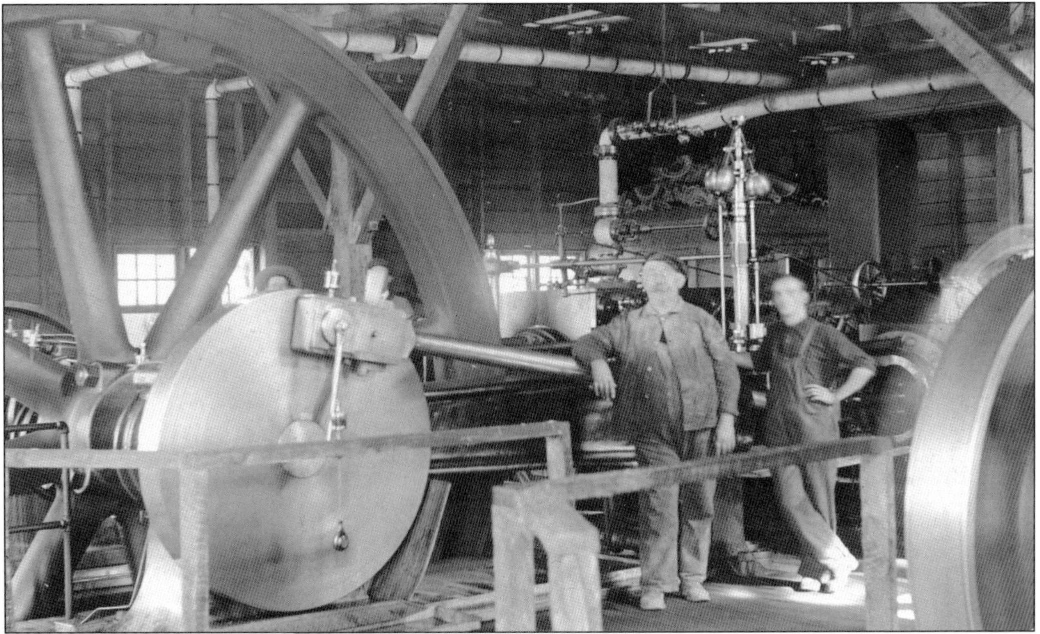

These workers assured the power plant was operating continuously. Shown in this image are the massive flywheels that convert lateral motion into rotary motion for the generators. All large equipment came to the site by way of the U&D railroad. Unfortunately, due to a fire years ago, no U&D records remain. (Courtesy of David Turner.)

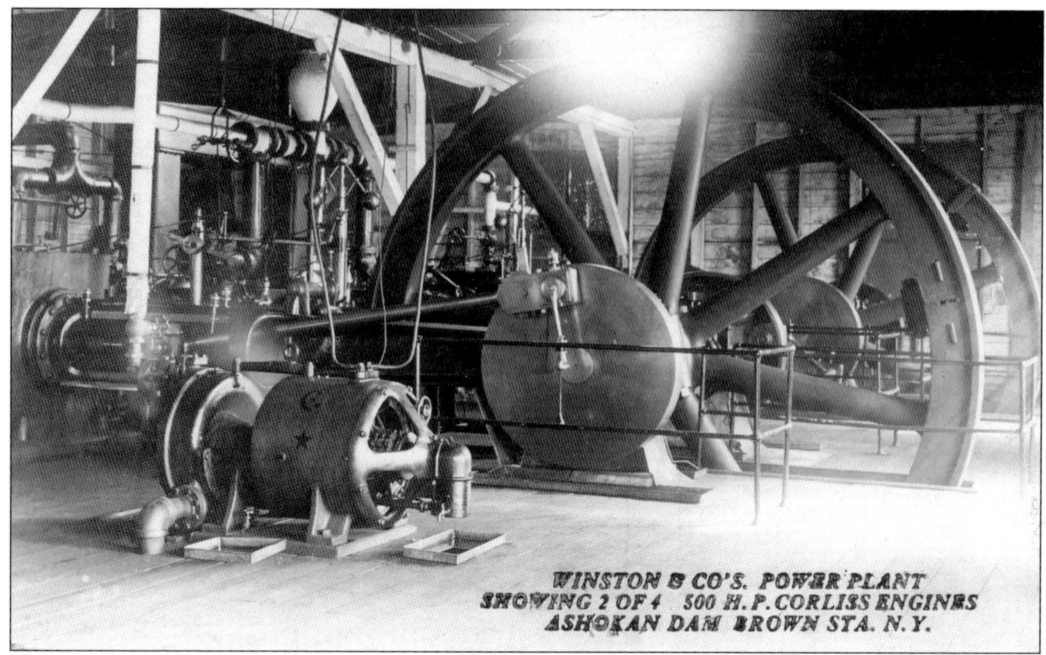

This view of the inside of the powerhouse shows one of the generators. It is difficult to determine, but with the large pipe connected to the cover at the front of the generator, it appears to be steam-driven. Steam was delivered at 80 psi, which was strong enough for lighting the camps or for night work.

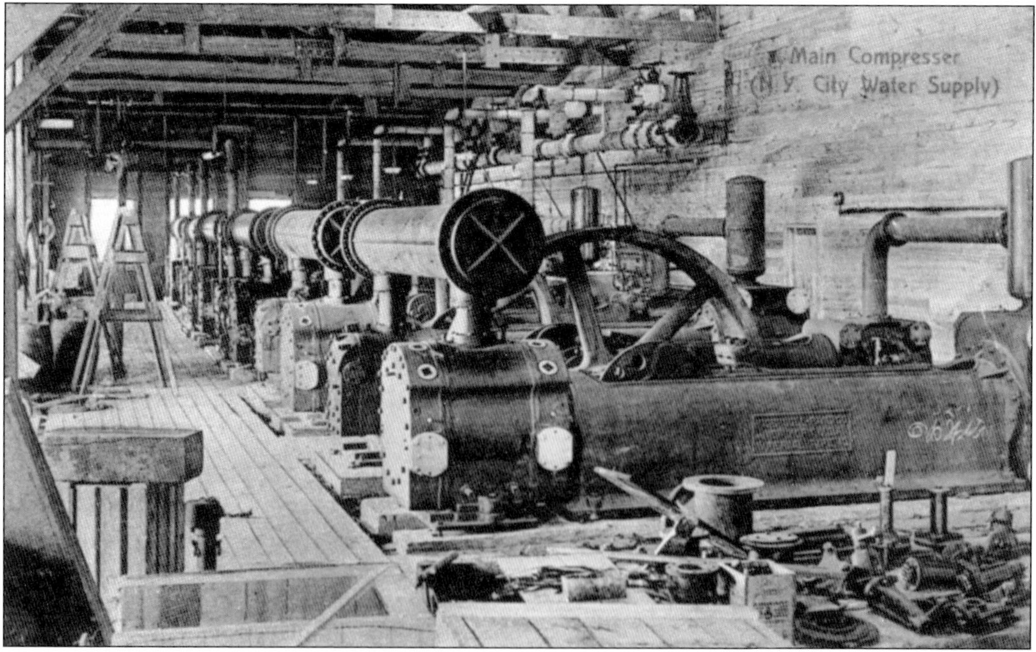

This is a view of the Ingersoll Rand air compressors, which could each deliver 3,500 cubic feet of air per minute. Even after a few miles of piping, the pressure was 80 psi. Piped air was used for rock drills in the cut-off trench and for tools at the far end of the dividing weir. The air piping was 4 to 12 inches in diameter.

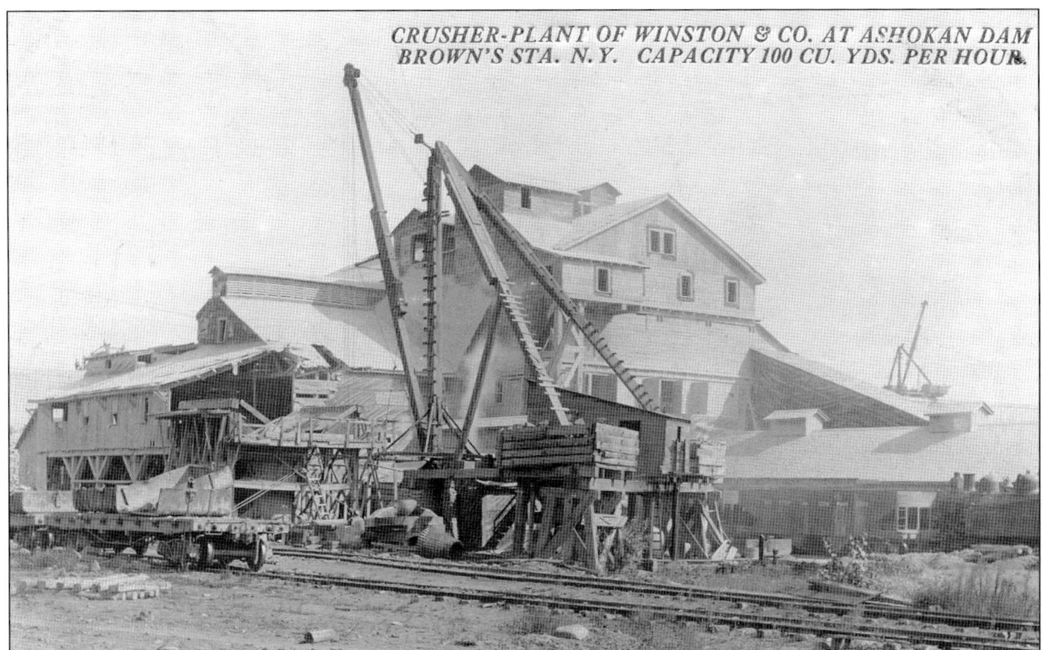

The rock crusher made aggregate for concrete. Five-yard skips of quarried rock were delivered on flat cars. The plant used a vertical steam-driven cone-design crusher set to produce pieces smaller than 2.25 inches. High-pressure steam from the power plant 150 feet away was delivered through a six-inch pipe to the crusher drive system. The entire site was a very noisy place.

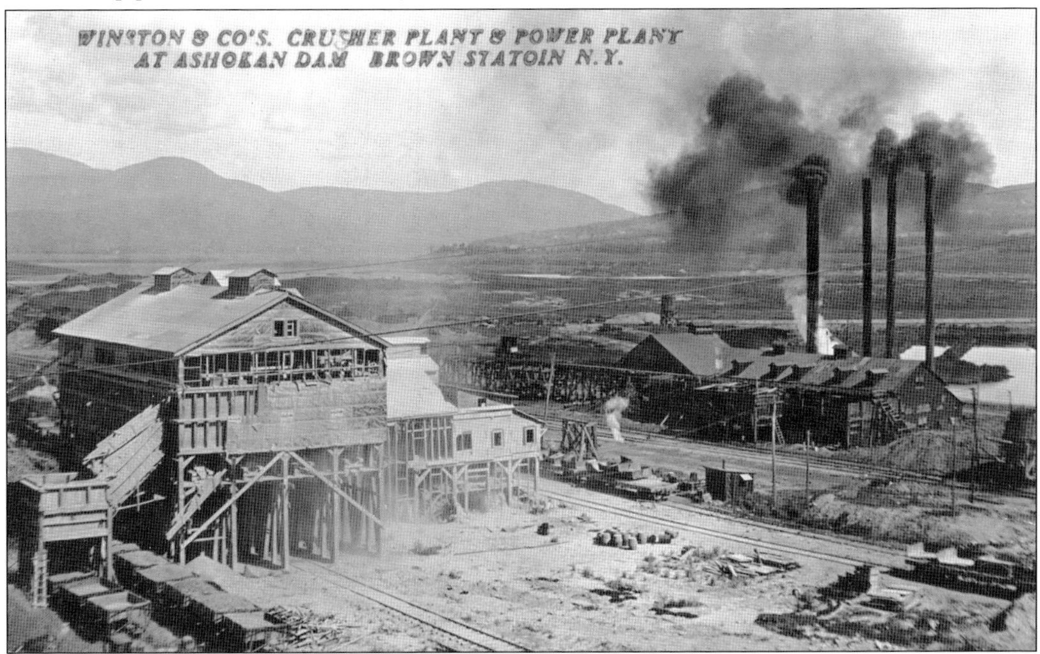

This image is of the opposite side of the crusher, where the crushed stone was loaded for the cement plant. The crusher used a heavy rotating steel cone with short vanes inside a similar cone. Rock fell into the top, where vanes trapped the stone between the inner and outer walls until it fractured and fell through to a continuous belt. Large pieces were sent through again.

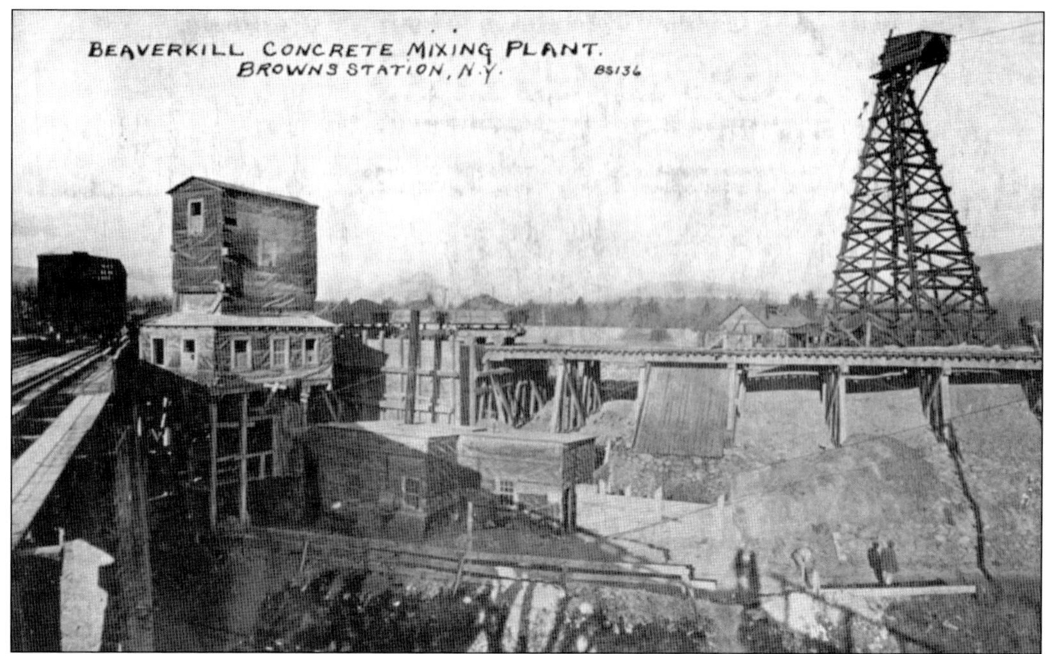

This postcard shows the concrete mixing supply yard. Cement was delivered in bags or barrels on the elevator at left. Aggregate and sand were delivered to hoppers by rail. Materials were moved out of this view to the mixers. From here, all of the concrete in the dam and dividing weir was mixed, beginning with all the cast block and mix for cyclopean masonry.

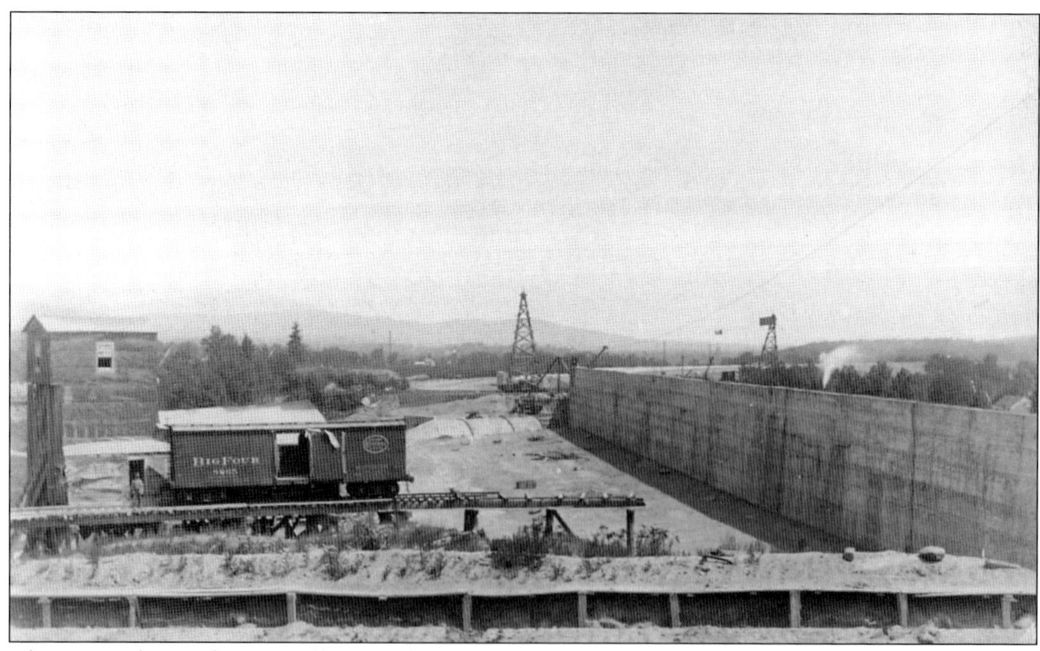

This view shows the partially complete West Dike core wall and a boxcar near the cement elevator. By October 19, 1911, a total of 1.1 million barrels of cement were unloaded here. In the background, the cover over the U&D dike embankment tunnel is visible. This photograph was taken before June 1913, as the railroad is still running through Brown Station.

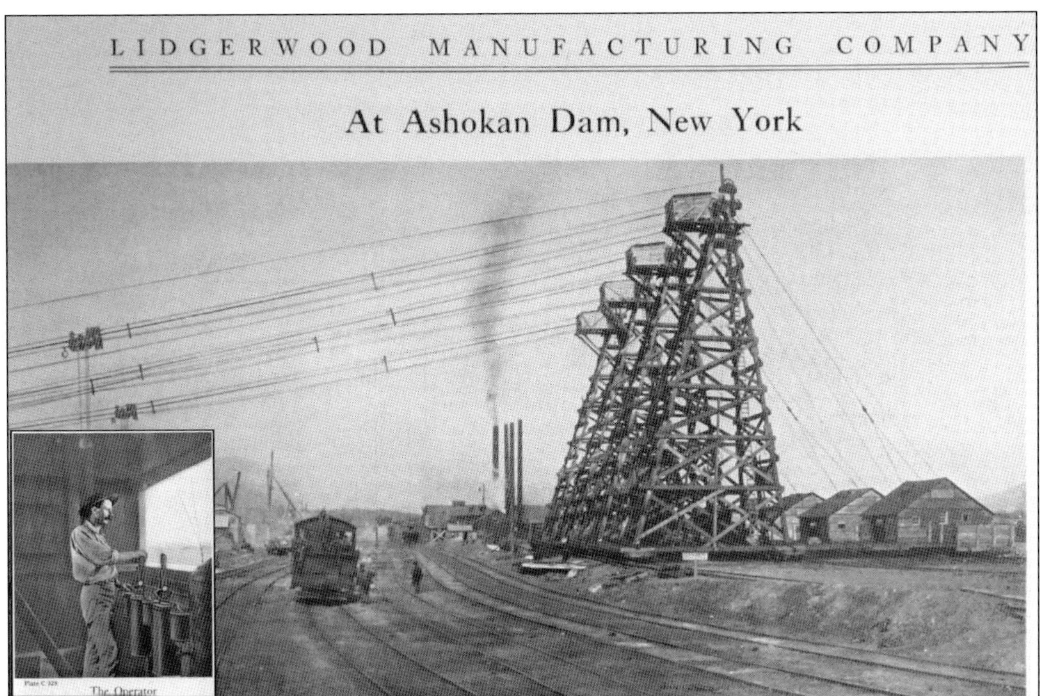

This Lidgerwood Manufacturing Company catalog page shows the Ashokan cableways. The cableways had a 1,534-foot clear span, could carry a 15-ton load, and moved a thousand feet per minute. The towers were on movable platforms with compressed air driving winch motors, and split cables anchored the tower to its platform. The operator in the tower controlled the function of the carriage. Cableways were used to relocate derricks.

The carriage rode the heavy cable; the top cable was for stabilization and load location, and the lower cable moved the carriage. A fourth cable lowered and raised the load. The tower operator was flagged to tell him when to lower the load. Accordion sections opened for precision location. The lower cable extended from a winch in the powerhouse. The button on the upper cable indicates a drop point.

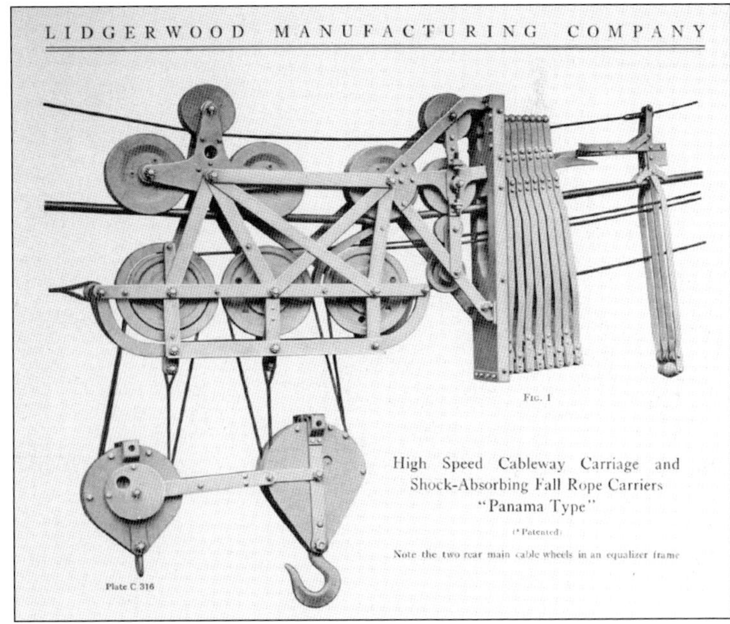

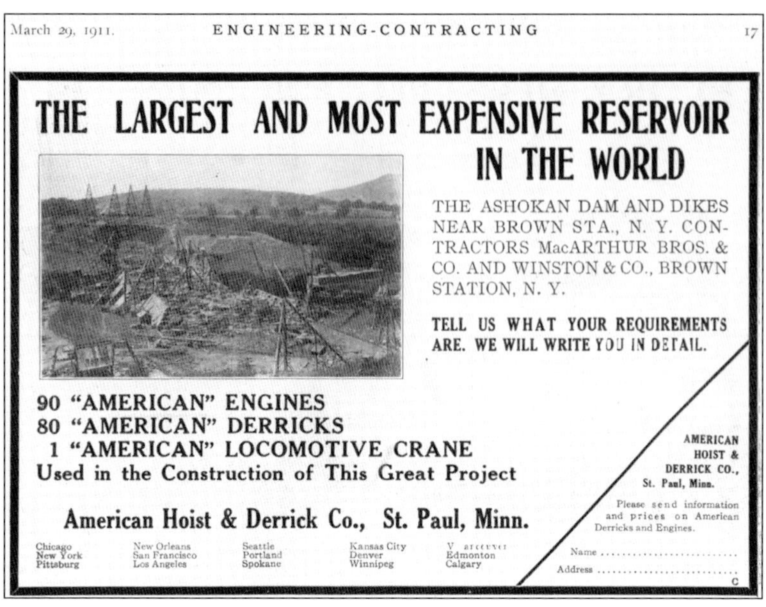

This ad for American Hoist & Derrick Co. from the March 1911 issue of *Engineering-Contracting* states what the reservoir contractors were using. Most of the engines listed in the ad were used with American derricks to lift stones and cast blocks for the masonry in the Olive Bridge dam. Other derricks were used for moving dike forms, pouring concrete, and other jobs.

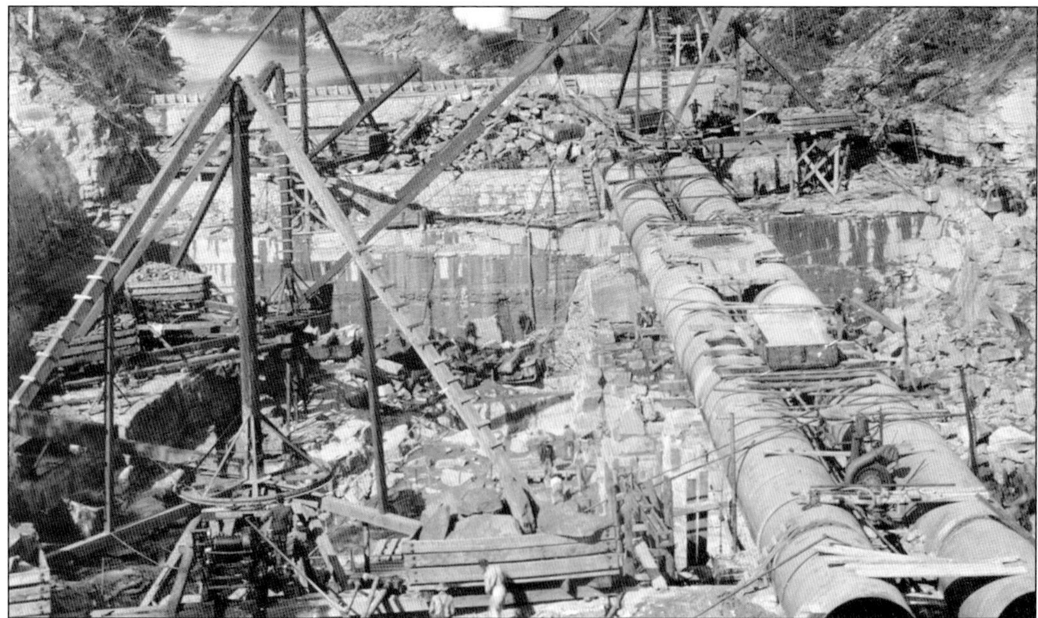

This image of dam foundation rock removal features a number of derricks. The closest one shows the hoisting engine at the bottom with two support legs. The right one is anchored in a stone-filled base to the right of the rotator. This derrick has its boom directly in line with the support beam. The derrick behind has its boom to the right with the lifting block near the end.

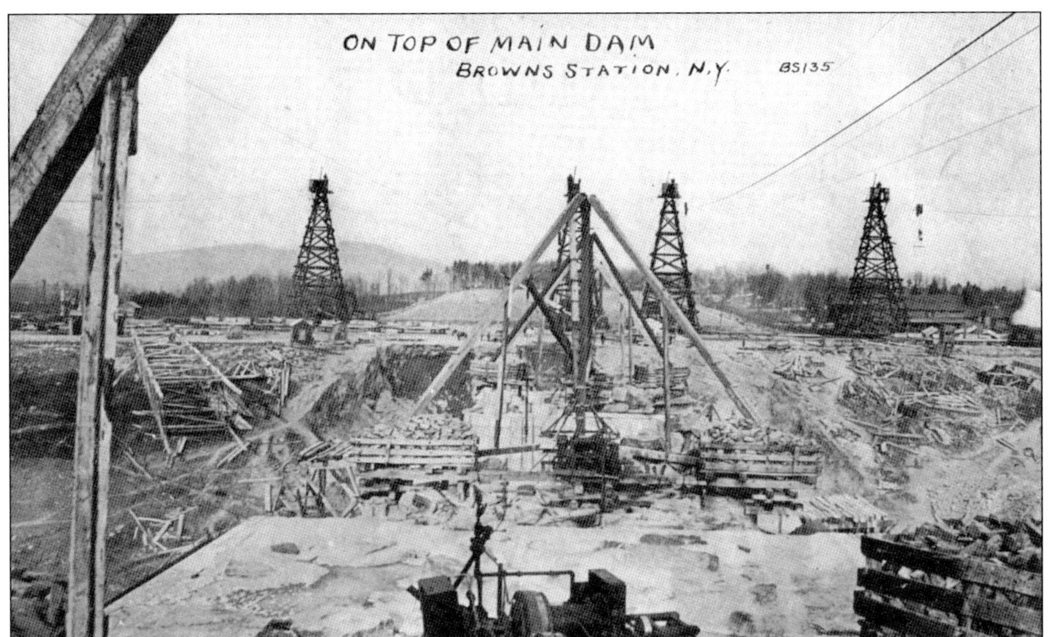

This image of the main dam looking north shows the hoisting engine of the derrick with the boom overhead and a good view of the next section derrick and its hoisting engine near the base. All these hoisting engines, as well as all others, ran on compressed air piped from the compressor plant.

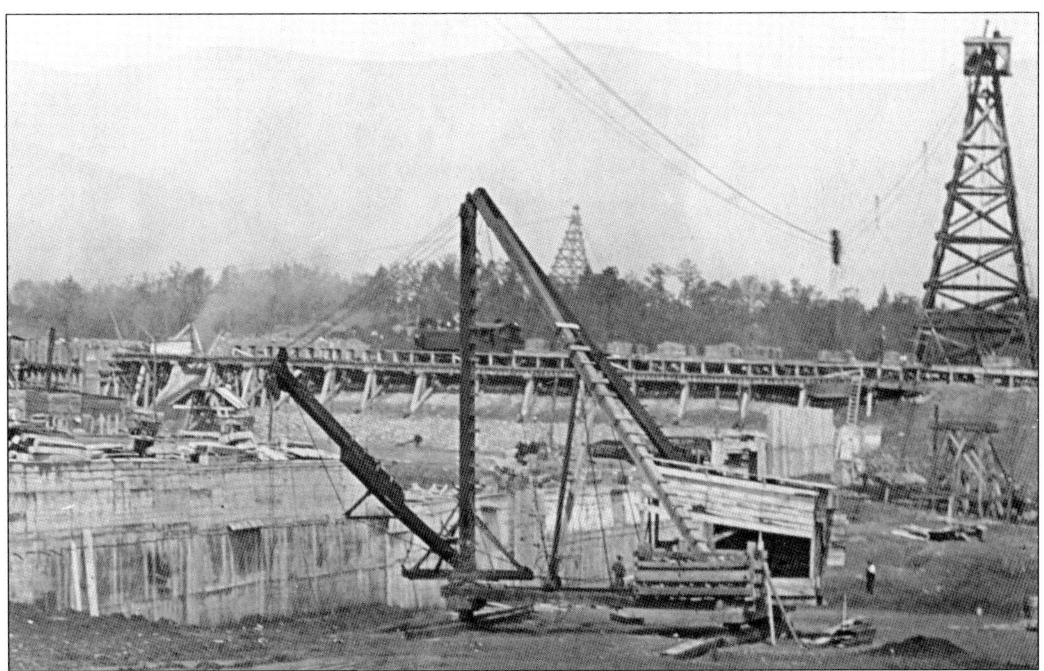

Work commences on the Beaverkill section of the Middle Dike, filling the excavation needed to block the pre-glacial gorge found during earth removal. The two rear derrick legs straddle the hoisting engine inside its enclosure. The leg bases were very heavy timbered boxes that appear empty, probably being set in position. The cableway has a skip shown under the black carriage.

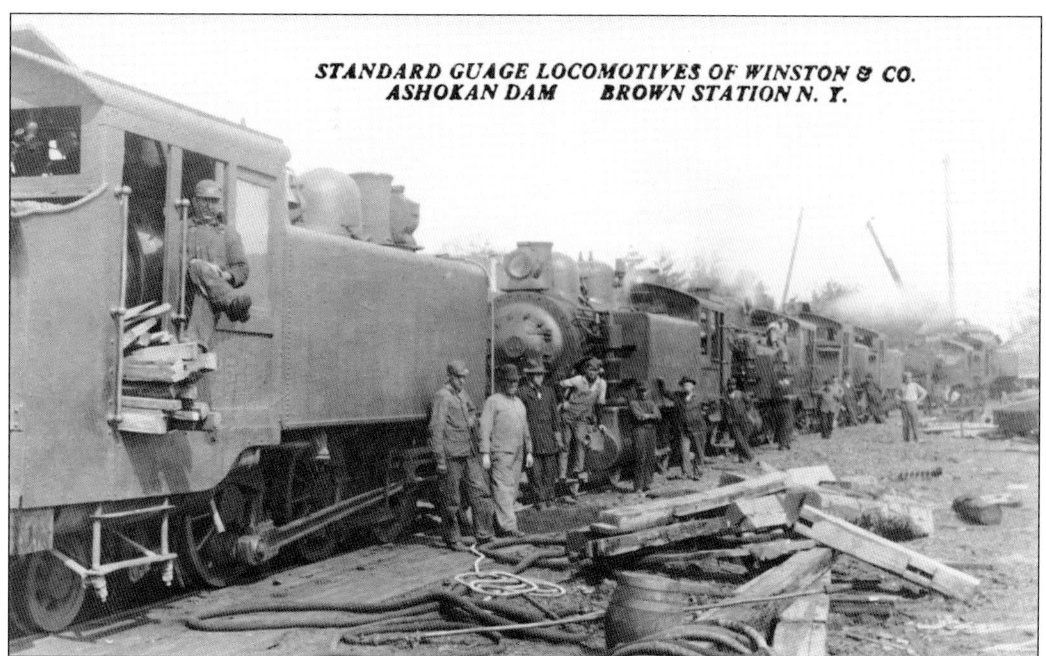

These locomotives are part of a group of nine 40- and 65-ton 2-6-2 configuration ALCO saddle-tank steam locomotives. They are designed for the kind of operation where a tender could block visibility when running in reverse. A small ALCO museum is being developed in the Schenectady area. The ALCO locomotive factory is now the site of Rivers Resort and Casino.

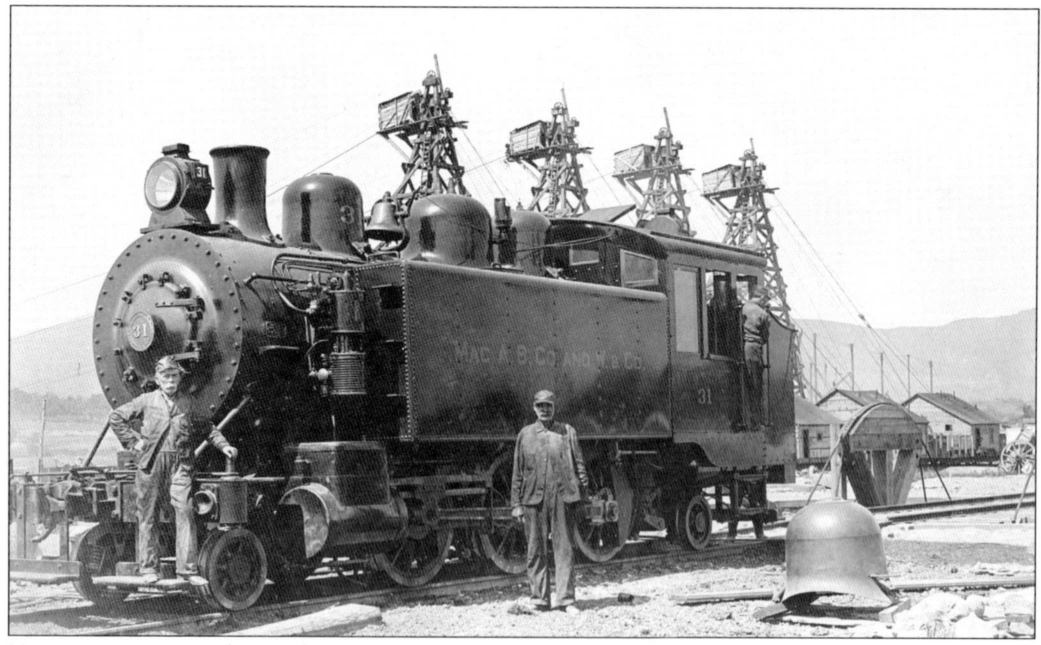

Here, a new 2-6-2 is being shown off. With six driving wheels, good motive power was available for hauling quarry stones or loads of embankment material. Note three domes on the top; two are sand for traction running in both directions. The cement plant was, and still is, located in Catskill, New York.

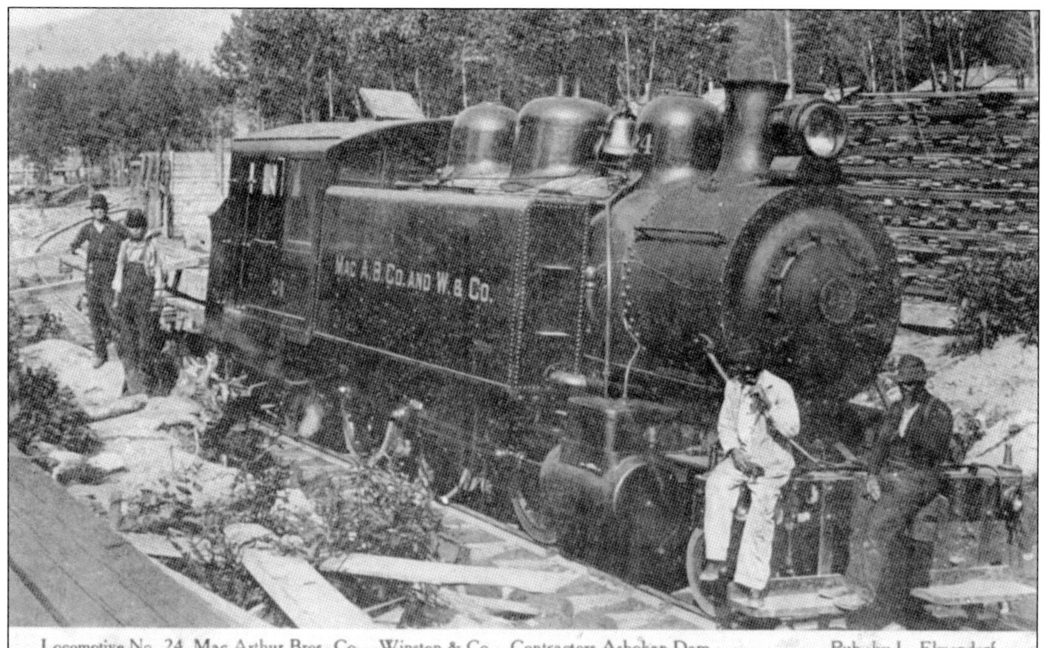

After work on the Kensico Dam, No. 24 was required to work the Ashokan site. Pulling cars loaded with key office and business material and many other items, it made its way across the Hudson River. The additional locomotives may have been coupled together and pulled by one engine, maybe following along with the No. 24 train. Since these engines had no fires, they were just dead weight and rolled easily.

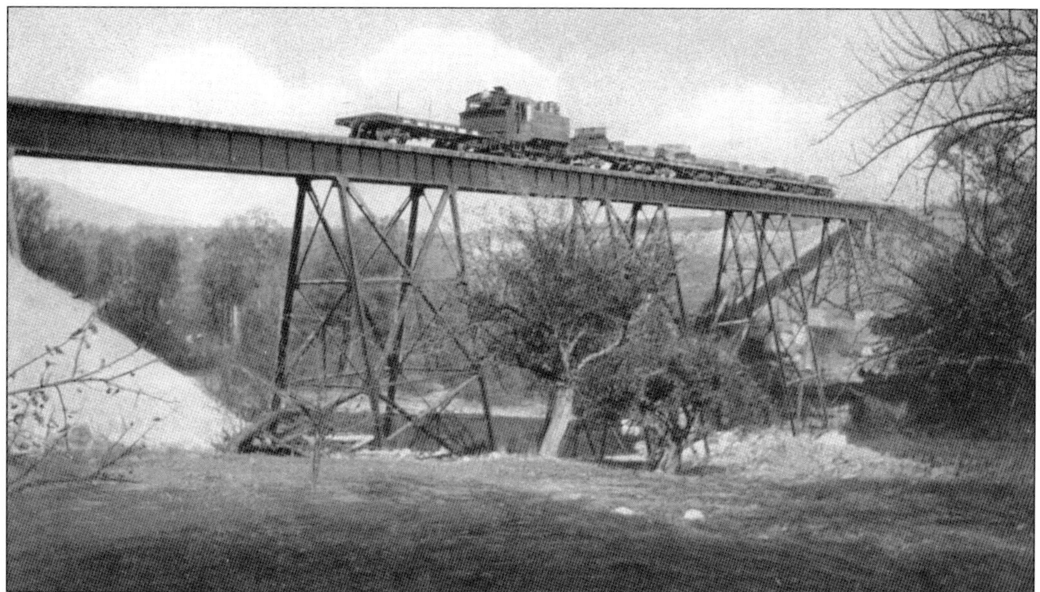

This trestle, 390 feet long and 85 feet above Esopus Creek, provided a route to the Yale Quarry. Much of the stone for aggregate, as well as the paving used on the waterside of the dike embankments, was hauled from the Yale Quarry. This trestle was damaged twice by floods. The tracks were finally moved to the Olive Bridge dam when it was completed, shortening the route to the quarry by two miles.

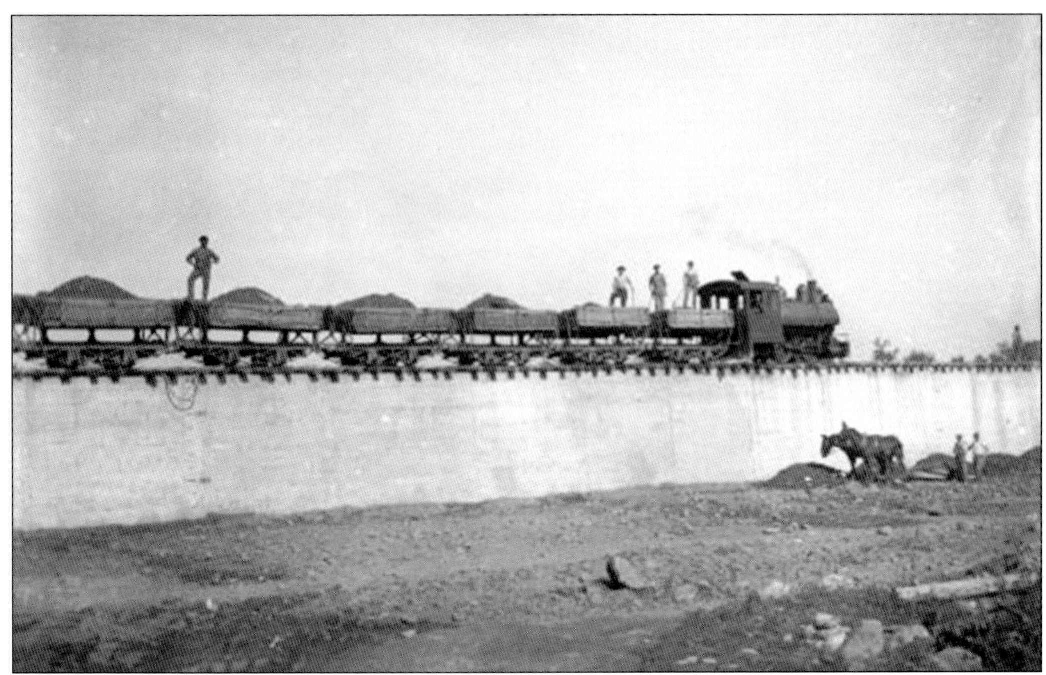

Locomotives have no value unless they are working. Here, carloads of earth are carried across a core wall to add embankment height. This narrow-gauge Porter locomotive is capable of pulling 12 to 15 cars that can dump to either side of the core wall. The contractor had 14 four-wheel Porters and over 250 four-yard dump cars. Most were used in the long period of embankment build.

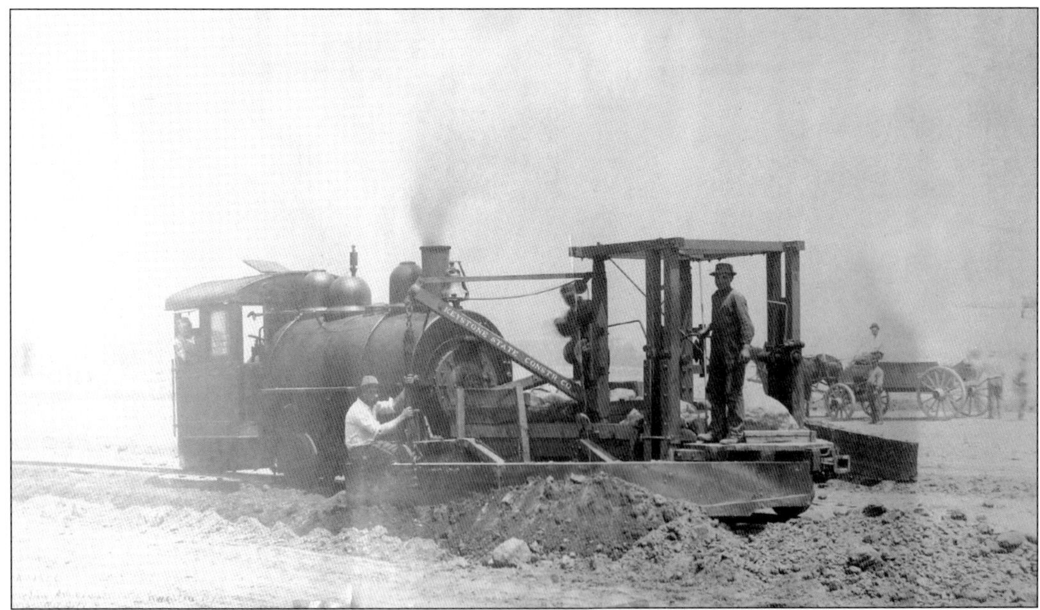

Another Porter locomotive pushes a grader to spread earth dropped for embankment fill. The apparatus works like a road grader but needs tracks close to its work. Note the box of rock used to counterweight the spreader blade. H.K. Porter locomotives were made in Pittsburgh, Pennsylvania. The firm closed after World War II.

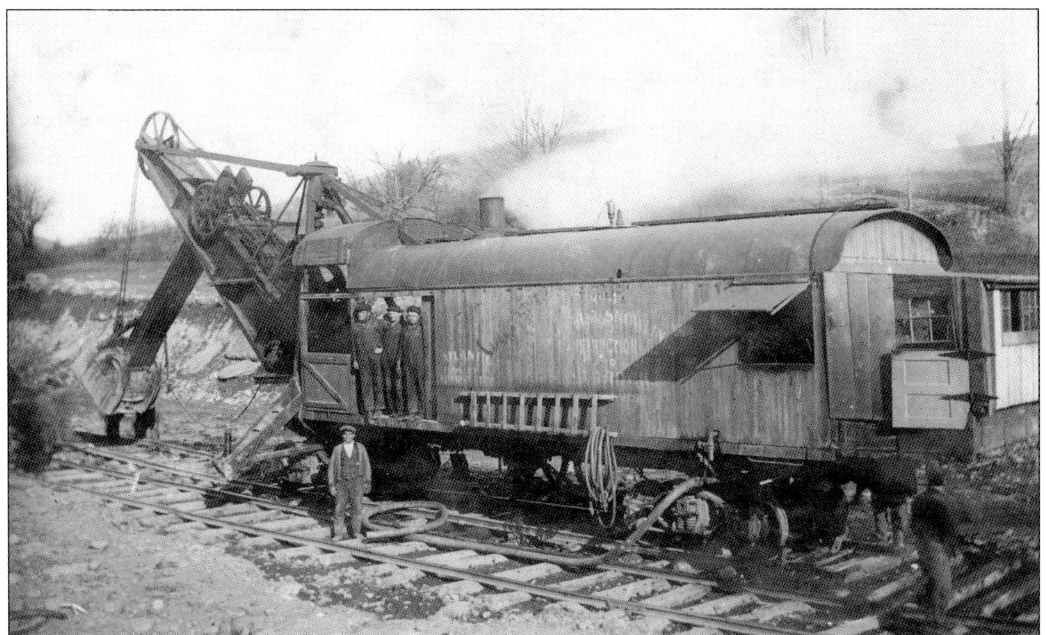

This vintage ALCO crane did a lot of heavy work. It was rated at 65 tons lifting load and was used mostly with one of the saddle-tank locomotives excavating earth along a rail line to load flat cars with skips for embankment work. It also loaded rock for the crusher with ease. This shovel was steam-powered through its onboard steam engine. Coal and water were carried inside the car.

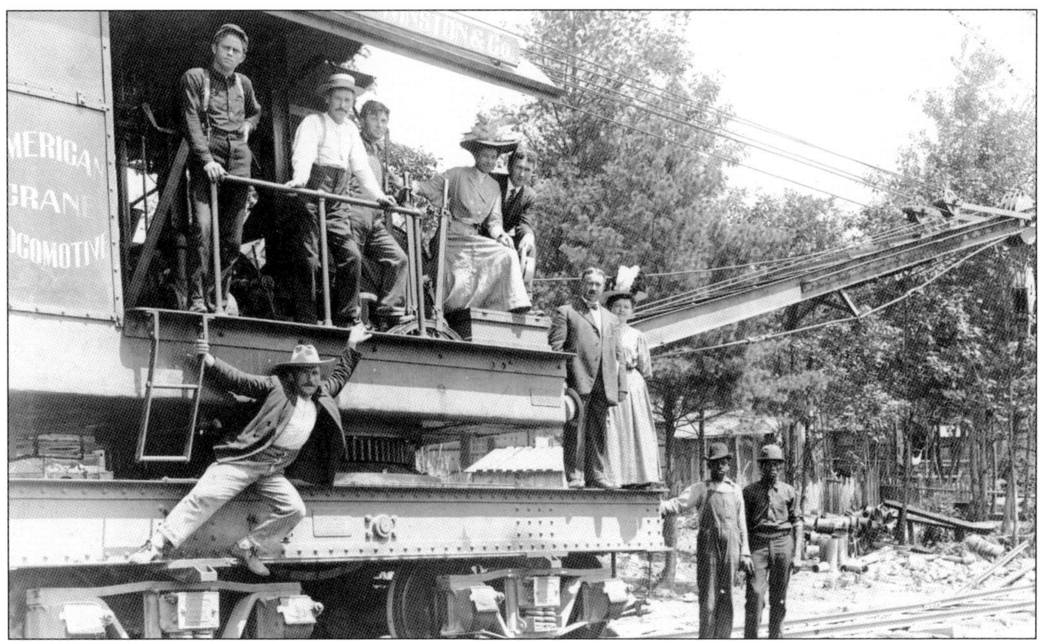

A new ALCO crane has just arrived at the Ashokan site. Pictured are contractors' principals with wives and crew. This is a different style crane than the one above. The cab is much more open. There is no indication that this crane was self-propelled. Today, ALCO is in operation under IPS cranes of St. Paul, Minnesota. (Courtesy of David Turner.)

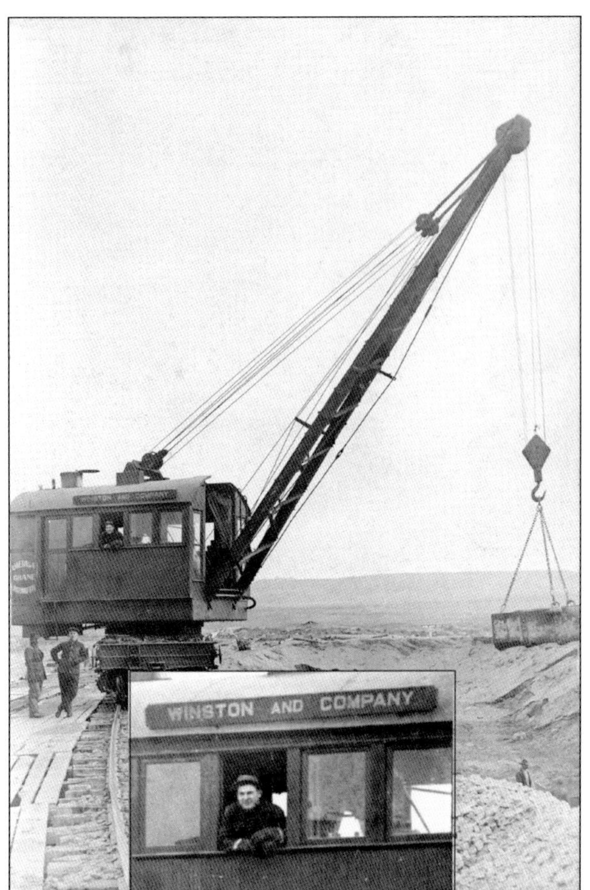

This long-reach ALCO crane was used to build and rebuild the trestle across the Esopus (see page 113), as well as many other chores that needed lifting without too much load. A heavy load with the boom low would cause this crane to fall toward the load. The signboard seen in the inset shows that the crane is part of Winston & Co.

This well-used steam shovel is similar to the railroad shovel at left but had to be moved by a traction engine. A coal-fired boiler supplied steam to the shovel actuators. The operator is moving earth for embankments from a borrow pile into a four-wheel bottom-drop Eagle wagon adjacent to the shovel's cab. Surveyors identified usable types of earth around the basin prior to the start of construction.

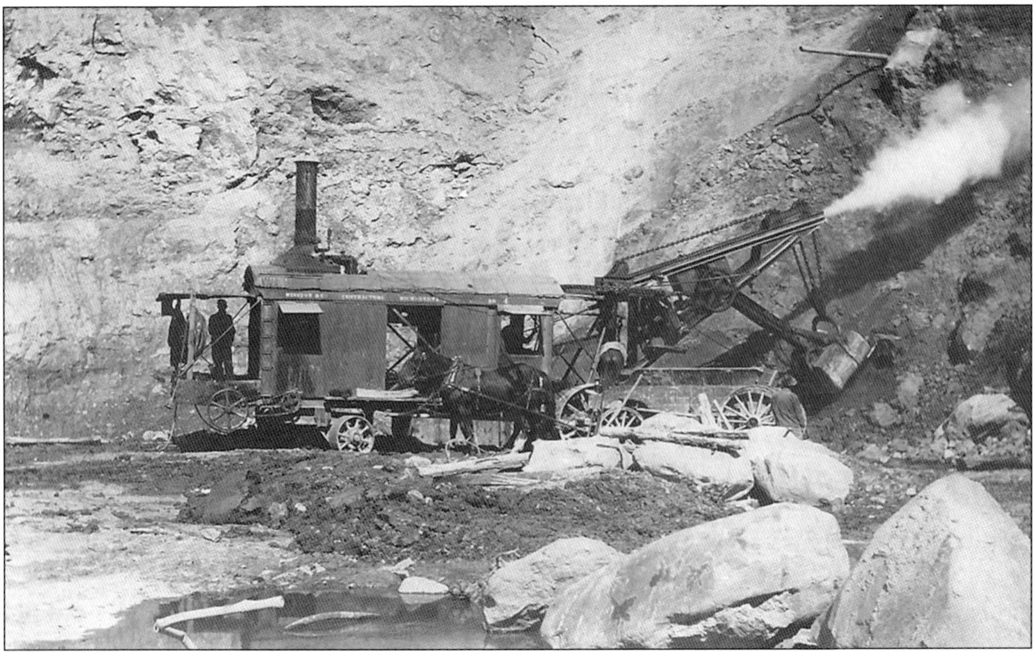

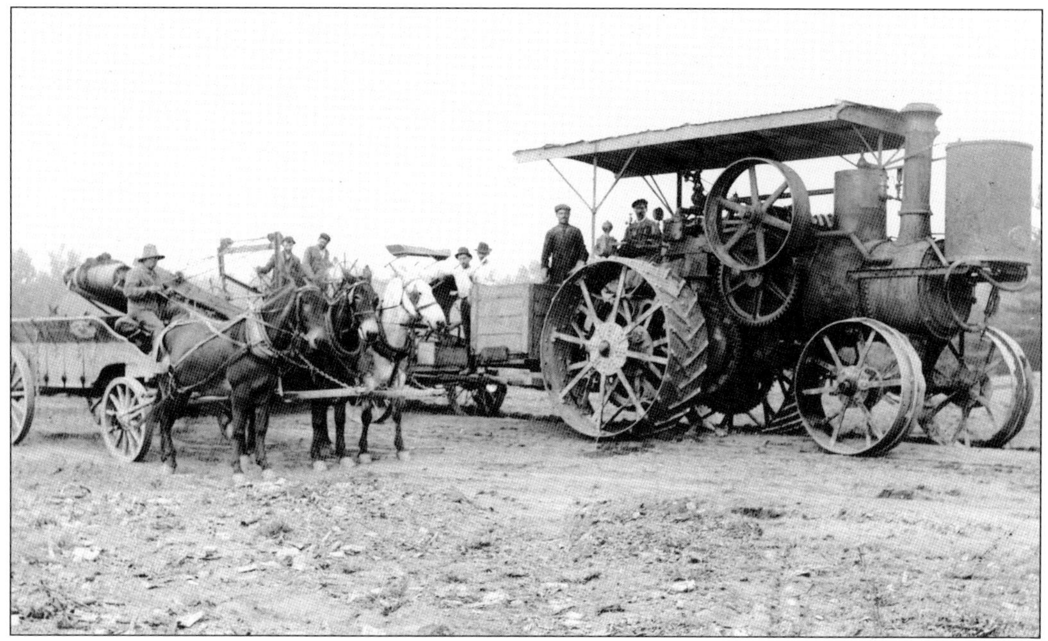

This steam-driven tractor, or traction engine, is designed to move heavy loads on construction sites. The tractor was made by Buffalo Pitts of Buffalo, New York, which was in receivership by 1918 due to the rise of the internal combustion engine. This tractor pulls an elevator for loading wagons. (Courtesy of David Turner.)

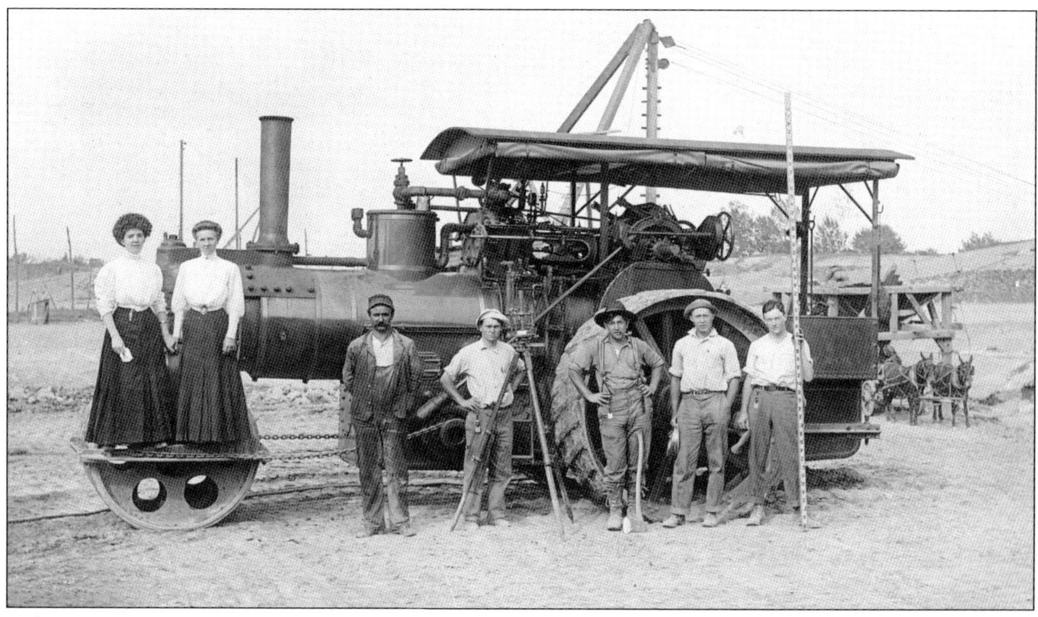

When Winston & Co. needed another compactor similar to the one shown on the top of page 84, it bought another. Here, the machine is shown delivered with a nearby survey crew and two ladies standing on the chain-driven steering bars. The steam-powered machine was used by driving forward and back, with about one-third overlap. (Courtesy of David Turner.)

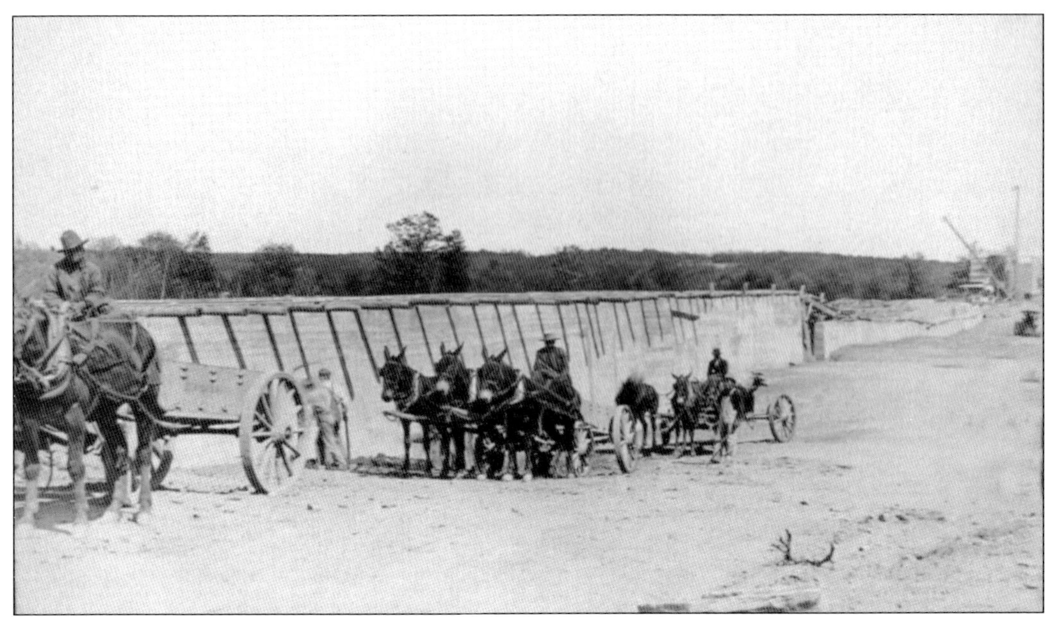

Mules and their drivers were invaluable. Here, they spread earth on an embankment with drop-bottom Eagle wagons. Earth was loaded by steam shovel, as shown earlier, or from an overhead skip. The drivers lined up to empty their wagons behind the previous load. Workers raked out rocks, and the compacting rollers flattened the base.

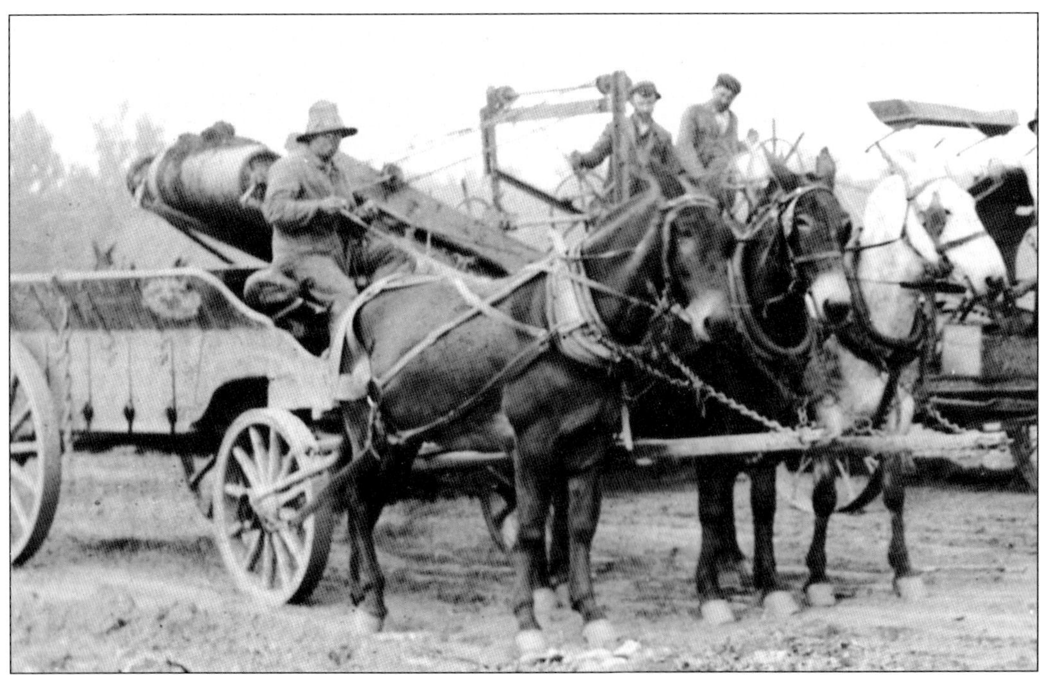

This detail from the image on the top of page 117 shows an Eagle wagon. The three latches on each side opened the bottom. The load spilled out as the team moved forward—just like a good dump truck. These men moved and spread many thousands of yards of embankment earth over more than six linear miles of dam wing and dikes. Horses are seldom teamed with mules.

Eight

It is Done

On July 19, 1914, all Winston & Co. steam whistles on the job blew steady for an hour. The major part of the work was complete. Only dike facing, some road work, and minor cleanup remained. The Ashokan Reservoir was sending potable water to New York City. The perimeter roads were serviceable, and everything was running as anticipated.

The work had been dangerous, with the possibility of being crushed or mangled quite high. For the seven years of construction, BWS records show there were 288 deaths and over 8,000 injuries across all projects. The reservoir project numbers were included; however, specific data is impossible to find. Since 1850, a state law froze access to coroner records. In *The Last of the Handmade Dams*, Bob Steuding writes, "some 40 Blacks and Italians have been killed on the job and buried in unknown graves." Surgery during this period was in its infancy. There is no question that workers with significant physical injuries would likely die. Amputation was the typical solution. Physicians at the time would do their best to stop bleeding, but most physicians' success rate was around 50 percent. The impact on residents was significant.

For many years on Memorial Day, valley residents would gather at the aerator basin. This no longer takes place, as very few remain with tales to tell. A few workers moved away, but most stayed in neighboring communities or moved into Kingston. Some residents moved by dismantling their home and rebuilding it in another location. Most often, homes were built to order. By the end of 1913, no one was living within the reservoir basin.

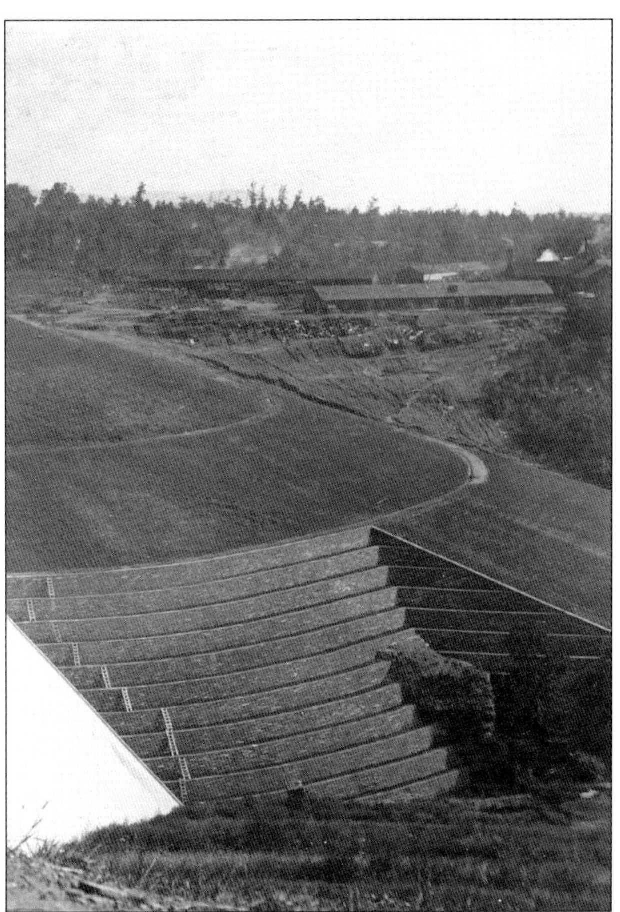

The toe of the dam is tied into the original Esopus Creek bed using a dry-laid bluestone terrace with seven-foot-high steps. This is the only place the sloping face of the 220-foot-high dam can be seen. In the distance are the lower main camp dorm buildings and two-story dining hall. More terracing is being done below the camp.

The Esopus Five Arches Bridge was the last to complete the reservoir roadway. All bridges were designed to be aesthetically pleasing. This bridge is due to be replaced in 2022–2023. About one mile in on Route 28A is the next bridge. New York State told New York City it was responsible "in perpetuity" to maintain the reservoir perimeter roadway. In 1913, the city tried to change the ruling but lost.

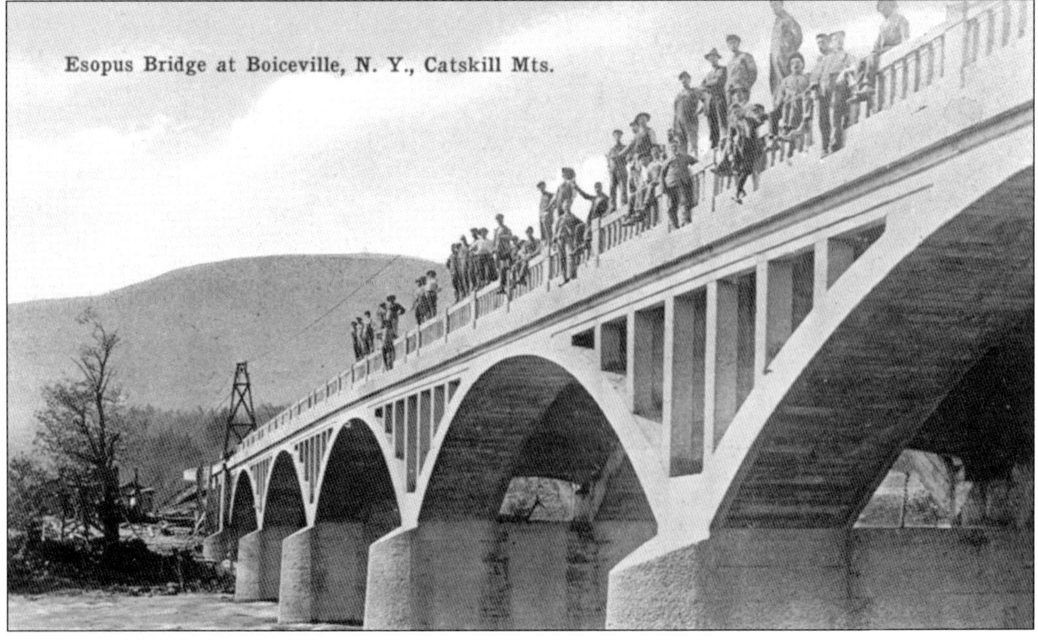

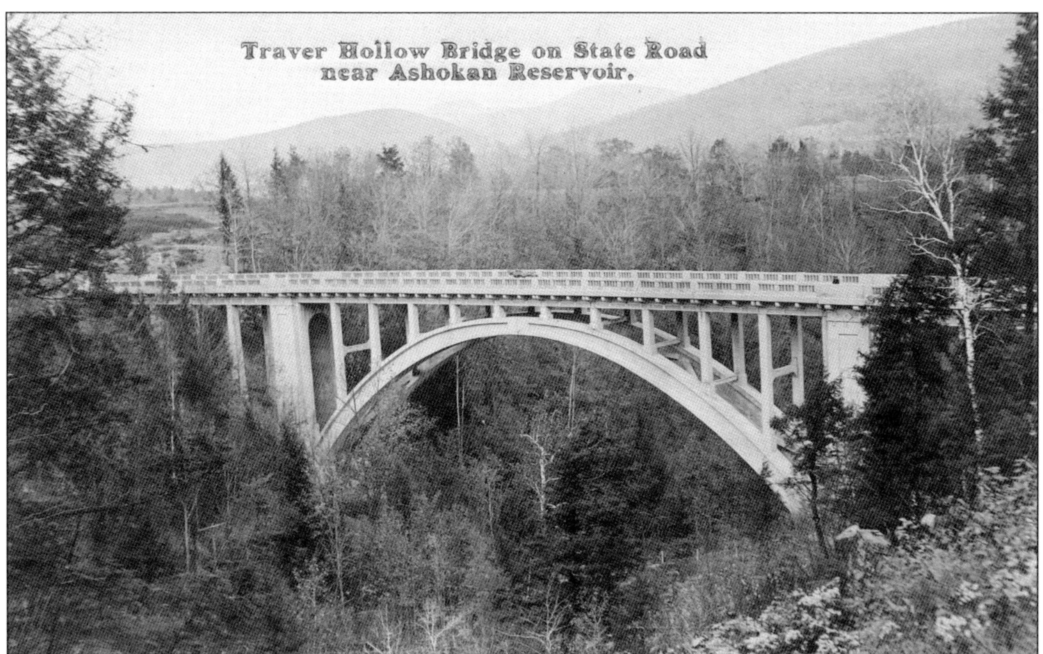

Completed in 1911, the Traver Hollow Bridge is approximately 70 feet above the stream bed. Traver Hollow Brook is another of the hundreds of glacial melt stream canyons visible throughout the Catskills. The bridge pictured was replaced in 1977 with a steel girder bridge. Residents who commuted across the bridge owned a second car and kept one on each side. A walkway was provided.

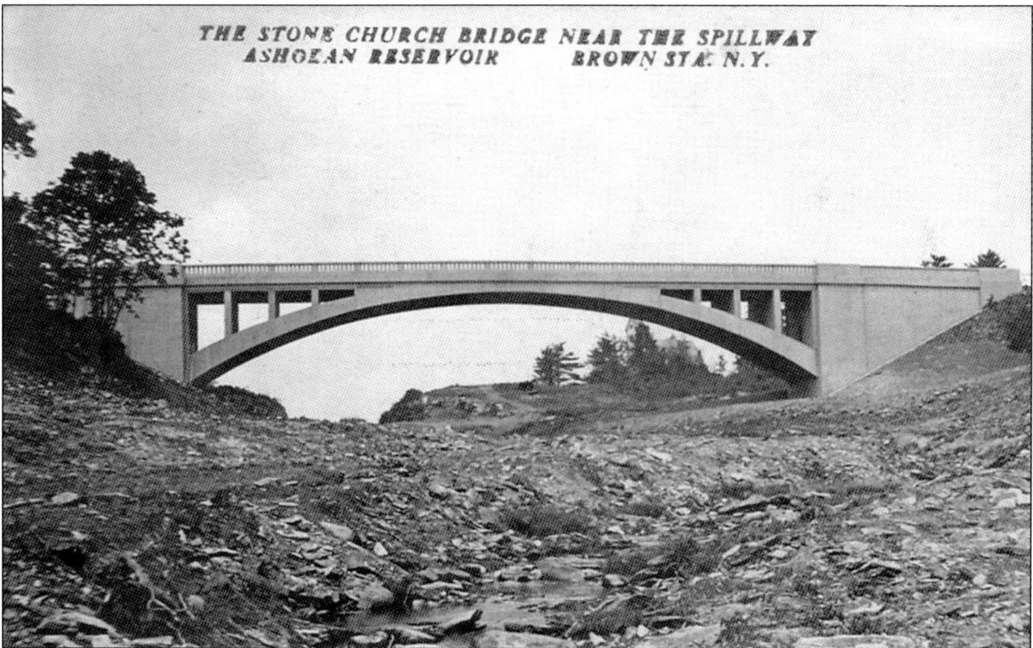

This bridge crosses the spillway outflow about a mile downstream of the spillway through remnants of another glacial canyon. The bridge had major parts replaced, but it is still similar-looking. The replacement was done to widen the roadway and increase load capacity. This bridge is fun to be near when the spillway is overflowing. Water flow through the narrow canyon is substantial.

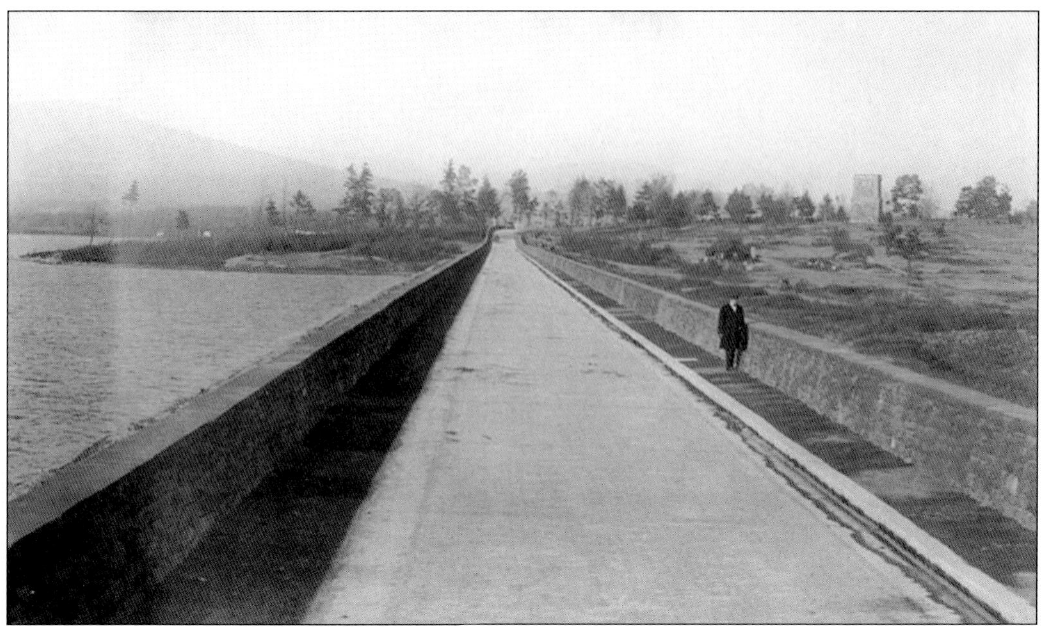

This view of the finished North Wing with Winchell Hill and McClellan's Tower on the right shows the dry-laid stone walls that once protected all the dike roadways. The bluestone walls were removed many years ago and replaced with standard guide rails to widen the roadway. The empty land to the right of the monument was the uphill edge of the main camp.

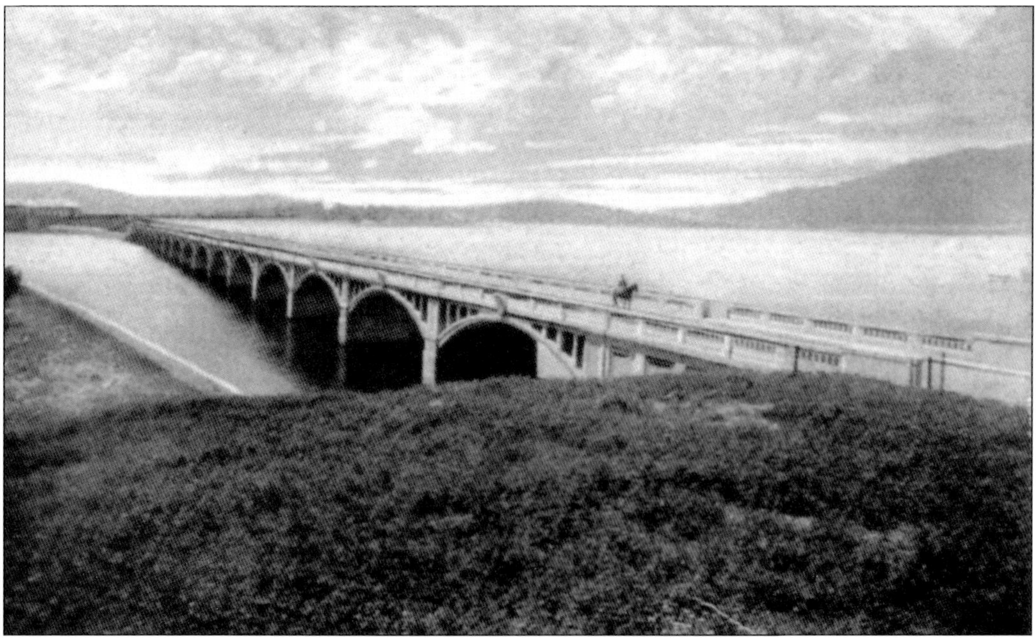

A lone rider, probably a BWS patrolman, is on the north end of the dividing weir not long after completion. This image is from a colored postcard and shows full East and West Basins. The first time the reservoir was full was on December 12, 1916. On November 22, 1915, the first clean water was released. All previous water was used to leak test the aqueduct during construction.

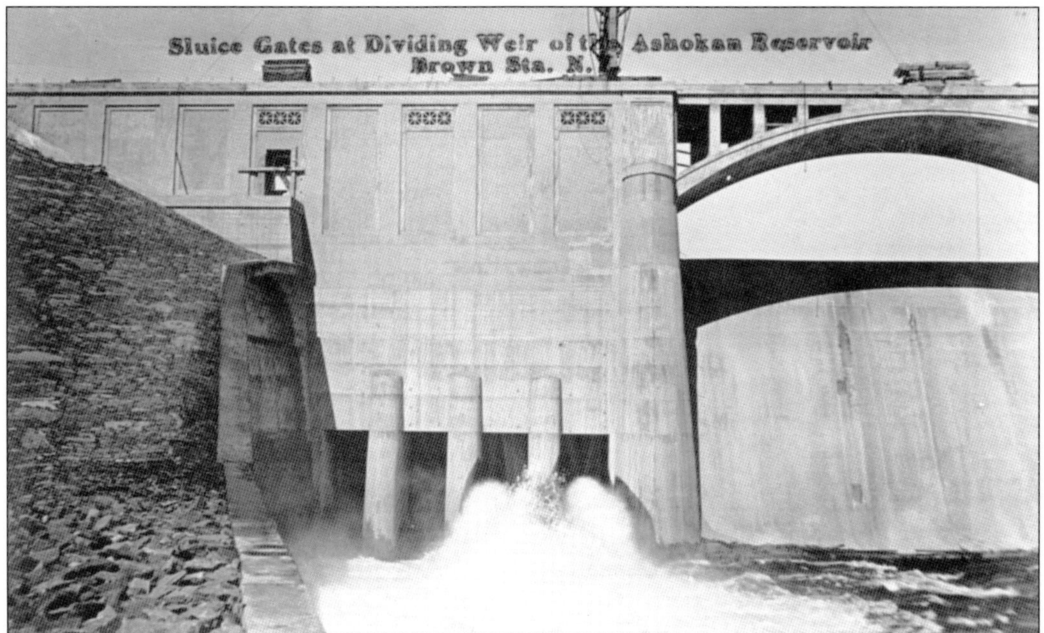

The dividing weir sluice gate chamber had two gates open when this photograph was taken. The photographer was down near the water. These gates are the only way to move water into the East Basin. In 2018, the New York City Department of Environmental Protection, which manages New York City water and wastewater systems, performed a gate repair. Only one gate change was covered in the local newspaper.

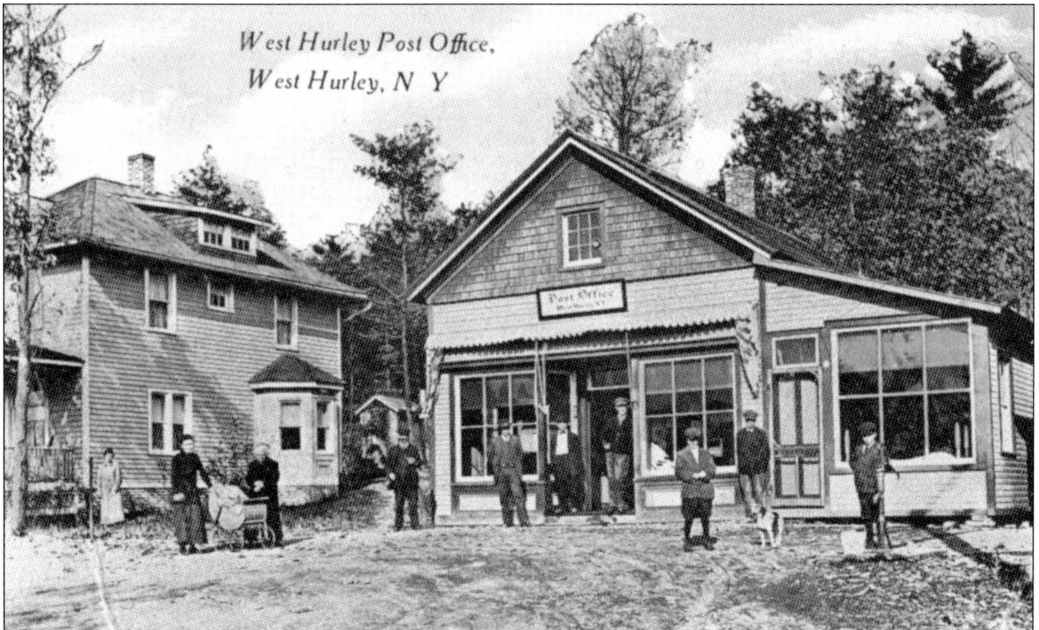

The hamlet of West Hurley moved north of its old location in the basin, at least the post office pictured here. This postcard was postmarked in 1917 from Spillway, New York, which was probably a temporary location for residents until a permanent building could be erected. This building still stands but is no longer a post office.

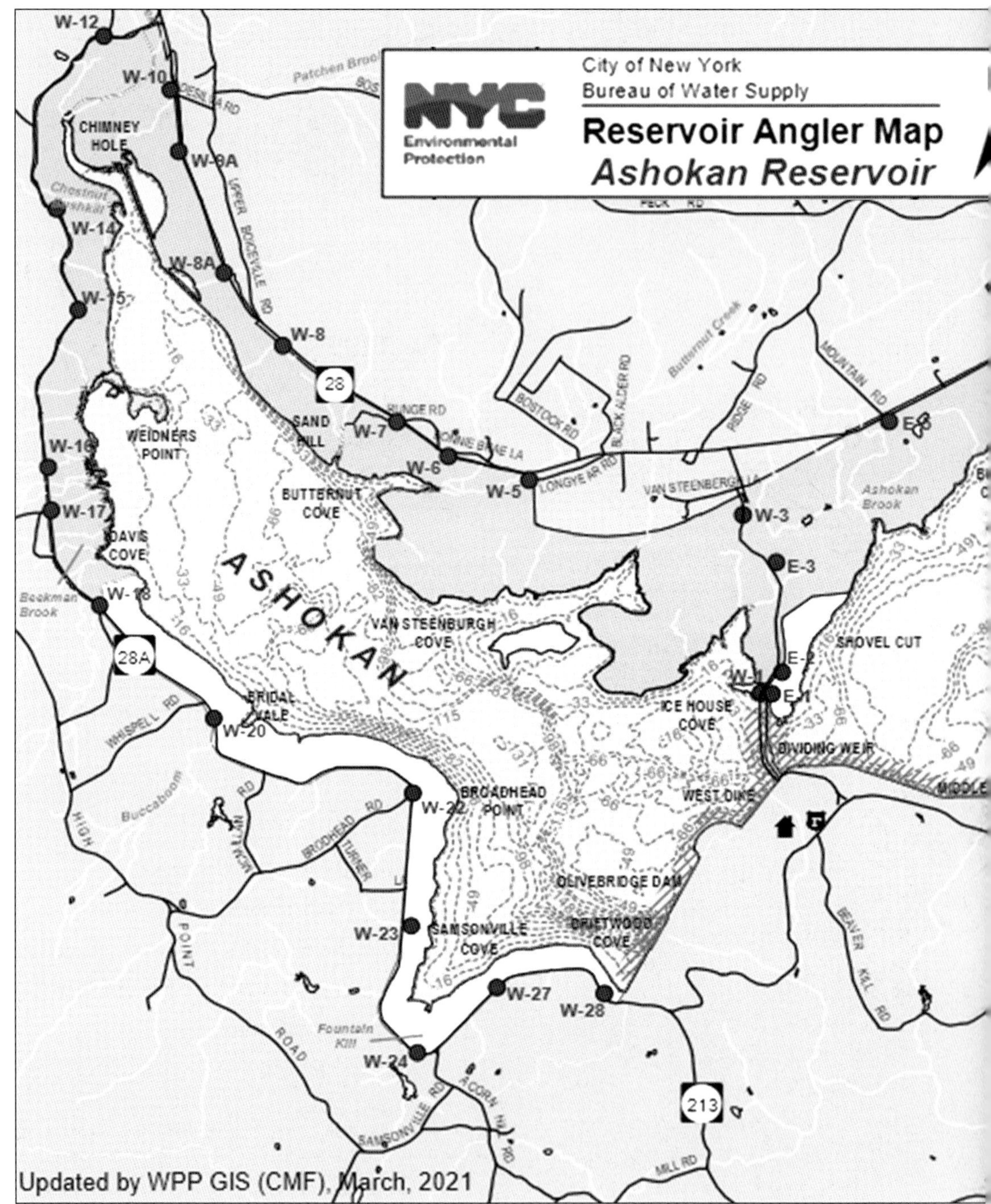

This anglers' map for the Ashokan Reservoir is a great asset for anyone wanting to fish there. The best feature of the map is the bottom contour. Looking at the dike locations and adjacent water depth, it is easy to see why they were needed to enclose the East Basin. The valley's east end naturally drained into the lower Esopus Creek as it flowed north to its mouth at Saugerties. Anglers must have an access permit, available online prior to entry, which must be carried while

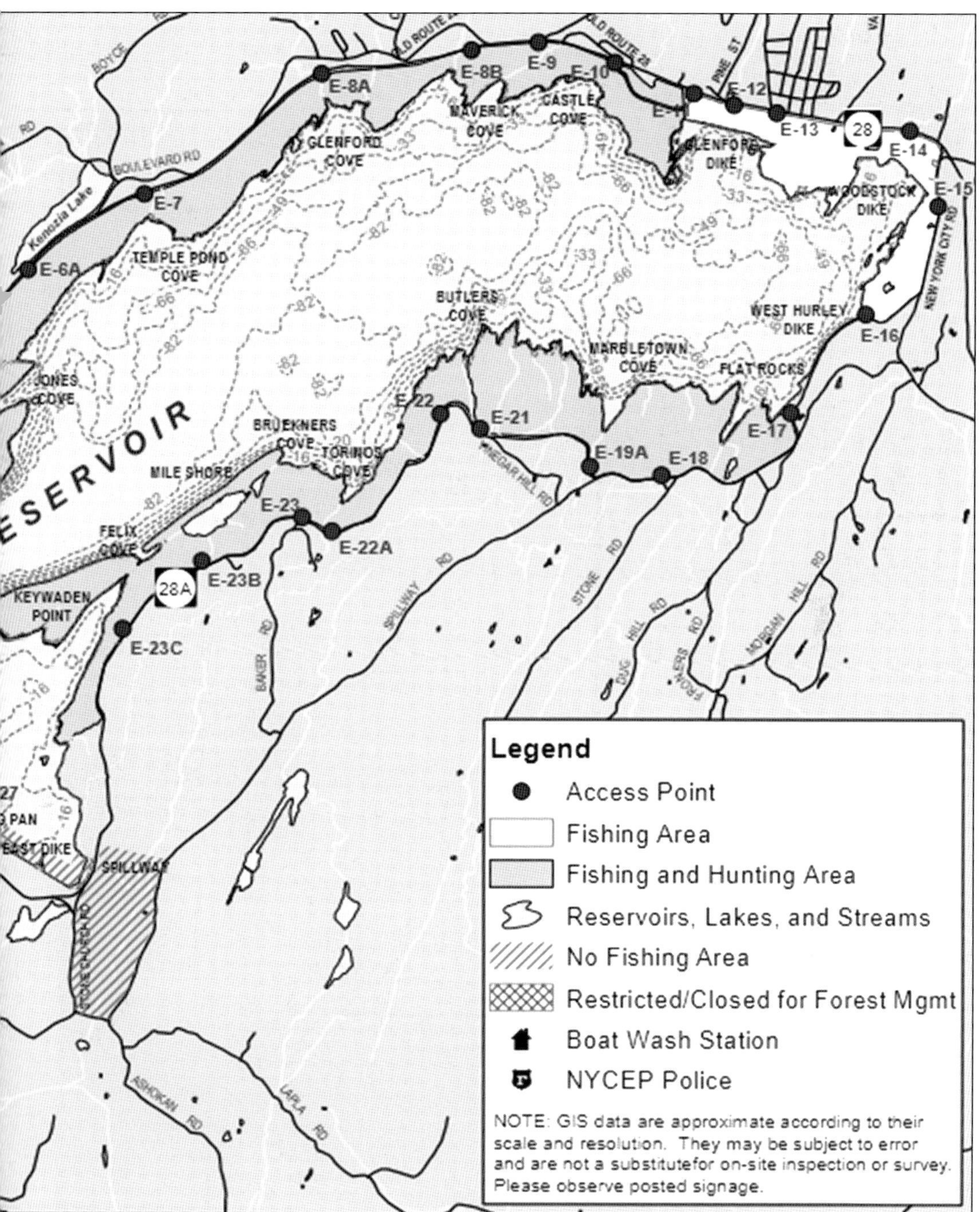

on the property. Almost the entire shoreline is open for fishing. There are access points along Routes 28 and 28A, gated and some with paths. The map is downloadable and capable of 500-times magnification. (Courtesy of the New York City Department of Environmental Protection.)

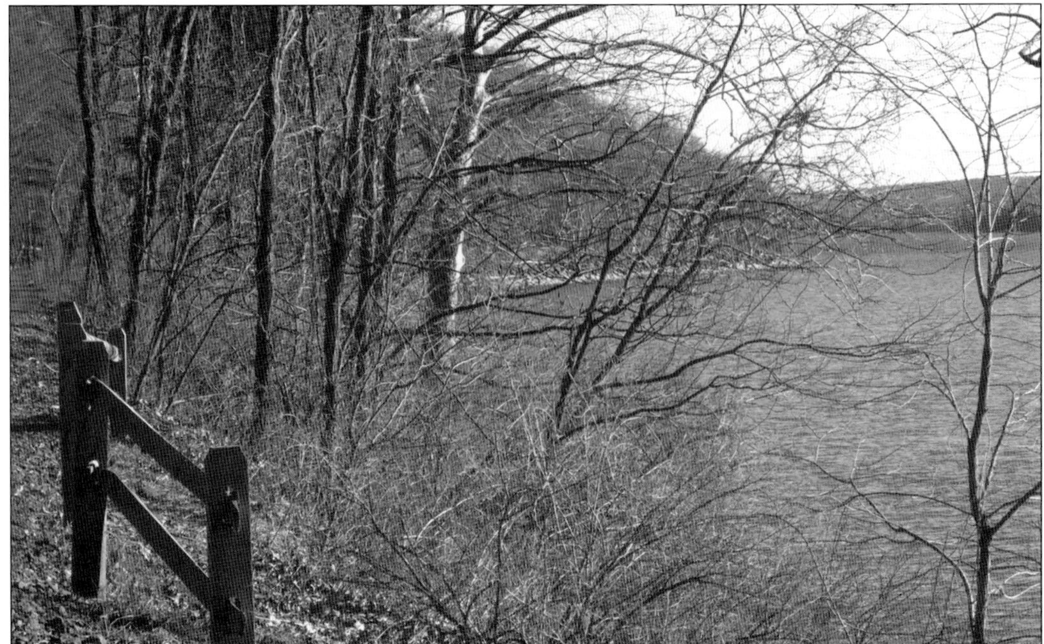

This image is from the new Ashokan Rail Trail on the abandoned U&D right-of-way, which has become a very popular trail. The photograph was taken in December 2020 in the afternoon from the north access off Route 28A. The trail runs the length of the reservoir, nearly 12 miles, but only the upper section has views like this. This view looks across the upper portion of the West Basin.

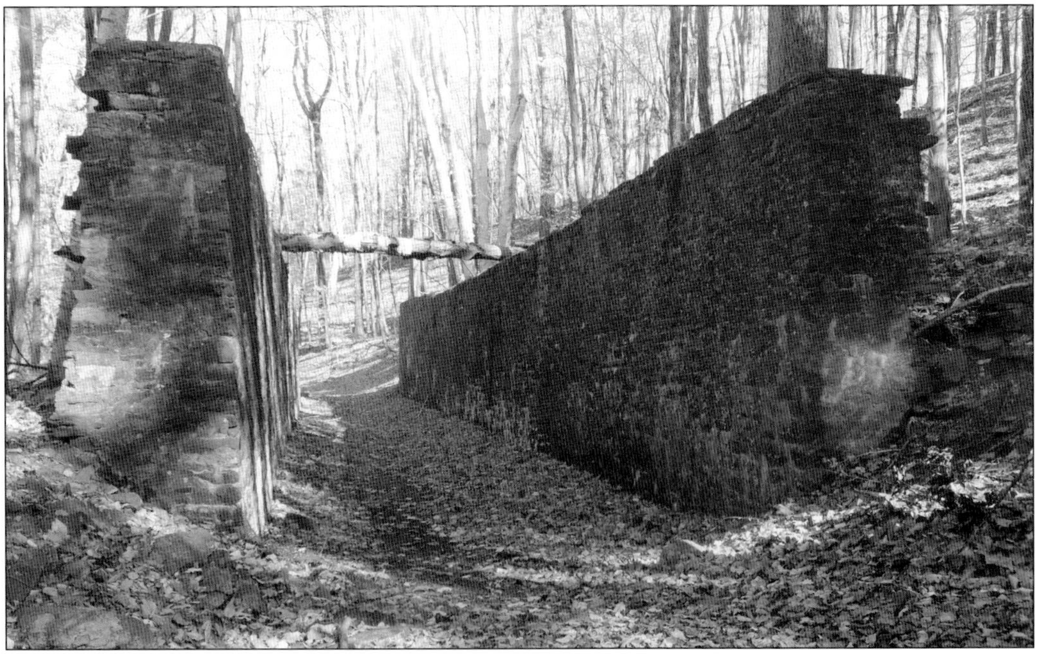

This dump car loading area at the Yale Quarry is where rock was crushed and loaded into six-yard side-dump cars that stopped between these walls. Loaded trains ran to the concrete plant, dropped the crushed rock, and then repeated the trip. The quarry also provided stone for dike paving and roadway walls. The area can be accessed via a one-mile trail from Route 28A.

Bibliography

Ashokan: Illustrated and Descriptive Account of the Main Dams and Dikes of the Ashokan Reservoir. Brown Station, NY: E.G. Nimsgern, 1909.

Bergner, C.S. "Construction Camp Life: Ashokan Reservoir." *York State Traditions*, Vol. 24, summer 1970.

Bone, Kevin, and Gina Pollara. *Water-Works: The Architecture and Engineering of the New York City Water Supply.* New York, NY: Monacelli Press, 2006.

Galusha, Diane. *Liquid Assets: A History of New York City's Water System.* Fleischmanns, NY: Purple Mountain Press, 1999.

Lidgerwood Cableways, Locks and Dams. New York, NY: Lidgerwood Manufacturing Co.

Light Locomotives. Pittsburgh, PA: H.K. Porter, 1906.

Newell, Helen-Marie. *The Hardhats.* Boston, MA: Houghton Mifflin, 1956.

Steuding, Bob. *The Last of the Handmade Dams: The Story of the Ashokan Reservoir.* Fleischmanns, NY: Purple Mountain Press, 1985.

Weidner, Charles H. *Water for a City.* New Brunswick, NJ: Rutgers University Press, 1974.

White, Lazarus. *The Catskill Water Supply of New York City.* New York, NY: John Wiley & Sons, 1913.